Praise for *Humans: A Brief History of How We F*cked It All Up*

"A laugh-along, worst-hits album for humanity. With the delicate touch of a scholar and the laugh-out-loud chops of a comedian, Tom Phillips shows how our species has been messing things up ever since we evolved from apes and came down from the trees some 4 million years ago."

—**Steve Brusatte, University of Edinburgh paleontologist and *New York Times* bestselling author of *The Rise and Fall of the Dinosaurs***

"Tom Phillips has proven beyond a doubt that humans are goddamn lucky to be here and are doing nearly nothing to remain relevant and viable as a species—except, that is, for writing witty, entertaining, and slightly-distressing-but-ultimately-endearing books about the same. And if you care to avoid orbiting the earth in a space-garbage prison of your fellow humans' design, you should probably read it."

—**Sarah Knight, *New York Times* bestselling author of *Get Your Sh*t Together***

"*Humans* is Tom Phillips's timely, irreverent gallop through thousands of years of human stupidity. Every time you begin to find our foolishness bizarrely comforting, Phillips adds another kick in the ribs. Beneath all this book's laughter is a serious question: where does so much serial stupidity take us?"

—**Nicholas Griffin, author of *Ping-Pong Diplomacy: The Secret History Behind the Game That Changed the World***

"Tom Phillips is a very clever, very funny man, and it shows. If *Sapiens* was a testament to human sophistication, this history of failure cheerfully reminds us that humans are mostly idiots."

—**Greg Jenner, author of *A Million Years in a Day***

"Chronicles humanity's myriad follies down the ages with malicious glee and much wit…a rib-tickling page-turner."

—***Business Standard***

Also by Tom Phillips

Humans: A Brief History of How We F*cked It All Up

TRUTH:
A BRIEF
HISTORY
OF
TOTAL
BULLS██T

Tom Phillips

HANOVER
SQUARE
PRESS

HANOVER
SQUARE
PRESS™

Recycling programs
for this product may
not exist in your area.

ISBN-13: 978-1-335-98376-3

Truth: A Brief History of Total Bullsh*t

Copyright © 2020 by Tom Phillips

First published in 2019 in the United Kingdom by Wildfire, an imprint of Headline Publishing Group.

This edition published by arrangement with Harlequin Books S.A.

Library of Congress Cataloging-in-Publication Data has been applied for.

HanoverSqPress.com
BookClubbish.com

Printed in U.S.A.

To my parents, who always taught me the value of truth. Although just FYI, I worked out that you were the tooth fairy. Santa's going to be so angry when he finds out you lied.

CONTENTS

"The most striking contradiction of our civilization is the fundamental reverence for truth which we profess and the thorough-going disregard for it which we practice."

Vilhjalmur Stefansson, *Adventures in Error*, 1936

AUTHOR'S NOTE

This is a book about things that aren't true. For fairly obvious reasons, this has meant that I've spent the past year in a state of almost permanent anxiety.

The book deals with history, and history is messy enough at the best of times, filled with provisional truths and half-truths and outright myths. In my previous book, which was about failure, I wrote that "the chance of this book about fuck-ups not including any fuck-ups in it is, frankly, minimal." (And yep: we have since found a few, thankfully none of them especially awful.) If writing about the topic of failure seemed like tempting the Gods of Fate, then choosing falsehood as the follow-up topic is basically presenting the gods with an open goal. And let's be honest, the Gods of Fate are unlikely to miss a tap-in to an empty net from two yards out. Not with the kind of form they're in right now.

So yes, there will undoubtedly be some mistakes somewhere in this book. I've done my best to avoid them: double and triple checking, going back to original documents wherever possible, trying to avoid the traps of overinterpretation. The endnotes should help you to check the facts yourself (and I'd encourage you to do so). But still, something will have crept through. Errors are inevitable; all we can do is try to minimize them, admit them, and mitigate them. That's one of the main points of the book! To that end, if you do spot a factual error—no matter how small—please email me at truth@tom-phillips.com. I'll be keeping a running public list of corrections at tom-phillips.com/mistakes-and-regrets/

INTRODUCTION

MOMENT OF TRUTH

You're full of shit.

Wait! Don't go. That was a terrible way to start a book—sorry.

I'm not really having a go at you in particular, here. That's especially true if you're browsing this book in a shop and wondering whether you should buy it. You should! You're very wise! Also, witty and stylish. To be clear, there's nothing about you especially that marks you out as being unusually untrustworthy or particularly given to falsehoods. (Unless you actually happen to be a professional con artist, I guess? In which case: Hi! You might enjoy chapter 4.)

You are, nonetheless, full of it: you're a liar, a bullshitter, and you're almost certainly wrong in hundreds of ways, large and small, about the world you live in. You shouldn't feel bad about that, though, because—here's the important point—so is everybody else around you. And, in the spirit of complete honesty, so am I.

What I'm saying is simply that, as humans, we spend our everyday lives swimming in a sea of nonsense, half-truths and outright falsehoods. We lie, and we are lied to. Our social lives rely on a steady stream of little white lies. We're routinely misled by politicians, the media, marketers and more, and the real problem with all of this is that it works; we are all suckers for a well-crafted fib. Perhaps the most pervasive lies of all are the ones we tell ourselves.

Right now, everywhere you look, you see dire warnings that we live in a "post-truth" age. Oxford Dictionaries crowned "post-truth" their Word of the Year in 2016; in 2017, no fewer than three books titled *Post-Truth* were published in the UK *on the same day*. Politicians seem to distort and spin and lie with increasing impunity. The public, we're confidently told, "have had enough of experts." The internet has turned our social lives into a misinformation battleground, one where we're increasingly unsure whether our Uncle Jeff is a real person or actually a Russian bot.

In fairness, it's pretty easy to see why people think we live in a uniquely fact-resistant time. To pick one rather obvious example: right now, the USA has a president who tells lies on a daily basis—or maybe they aren't even lies. Perhaps he simply doesn't know what's true and doesn't care to find out. The effect is roughly the same. According to the *Washington Post*'s fact-checking team, at the time of writing, President Trump had made 10,796 "false or misleading claims" in the 869 days since he took office,[1] following what they have described as "a year of unprecedented deception."[2]

That's an average of more than 10 untruths every single day, and if anything, the rate of his dishonesty seems to have been increasing as time goes by. He crossed the 5,000 fibs mark thanks to a particularly intense squall of bullshit on September 7, 2018, when he made no fewer than 125 false or misleading claims (according to the *Post*[3]) in a period of time totaling only 120 minutes. Which is more than one falsehood every minute.

That wasn't even his most dishonest day—that dubious crown is claimed by November 5, 2018, on the eve of the midterm elections, during which the *Post* recorded 139 inaccurate claims over the space of three campaign rallies.

This is, it's fair to say, not especially normal. But does it mean we're living in the age of post-truth? I'm here to say: nope.

Don't get me wrong. I'm not going to try to convince you that our present time *isn't* stuffed to bursting with a hundred thousand flavors of horseshit—it absolutely is! It's just there's a simple problem with the idea that we live in a "post-truth age": it would mean there was a "truth age" at some point that we can now be "post-" about.

And, unfortunately, the evidence for any such age is…uh… patchy, to say the least. The notion that we've recently left behind some sort of golden era of scrupulous honesty and a passionate devotion to accuracy and evidence is, to put it bluntly, a load of old baloney.

Yes, there's an awful lot of nonsense around these days. We all contribute to it in some way, whether it's large or small; we've all passed on an unfounded rumor, and we've all clicked that share or retweet button without checking the basics, because whatever it was appealed to our personal biases.

But, despite what you might have been told, we've been this way for a very, very long time.

That's what this book is about: the truth, and all of the ingenious ways throughout history that humanity has managed to avoid it. Because none of this is new. Donald Trump is very far from being the first politician to spray falsehoods in every direction like a fucked-up garden sprinkler. We've never needed a Facebook log-in to spread unverified and spurious rumors from person to person. For as long as there's been a quick buck to be made, and gullible people to make it from, there's always been someone willing to get creative with the facts in order to part people from their cash.

Of course, defining exactly what the truth is—and what it isn't—has never been as easy as some people might think. Then there are other questions, like...where does falsehood come from? Is dishonesty built in to humans and human society? Are humans the only creatures that lie? That's what we'll try to get our heads around in the first chapter, "The Origin of the Specious," where we'll explore the subtle differences between "lies" and "bullshit," discover the unexpected fact that there are different colors of lie beyond "white," and ponder the terrifying reality of just how many more ways there are of being wrong than being right.

For several centuries, the news industry has been one of our main sources of information about the world. Journalism, they say, is the first draft of history—but, as we'll see, it's often been a terrible first draft, the kind that has editors tearing their hair out. We'll look at the origins of our insatiable desire for news in chapter 2, "Old Fake News," where we'll meet a dead man who wasn't dead and discover that our modern anxieties about untrustworthy news sources and information overload are perhaps not quite as modern as we thought.

If the news business had humble origins, it didn't stay that way for long—it quickly expanded into an industry that shaped our societies and our view of the world in profound ways. That doesn't mean it got much more reliable, though. From the Great Moon Hoax of 1835 (when the New York *Sun* sparked a nationwide sensation with a series of entirely fabricated articles about how the famous astronomer Sir John Herschel had discovered a complex civilization living on the moon) to some complete bullshit about bathtubs, the Hitler Diaries and the infamous cat serial killer who stalked Croydon, a lot of what we've read about what's going on in the world has been total nonsense.

That's what we'll look at in the third chapter, "The Misinformation Age."

Not only have we been wrong about what's happening *in* the world; we've done a terrible job of getting anything right about the world itself. In chapter 4, "The Lie of the Land," we'll take a journey through several centuries of, uh, "creative geography." Whether it's vast mountain ranges that never existed, implausible tales of mythical lands, or explorers who may not have actually been to the places they claimed to have explored, we'll see how our maps have been shaped by the fact that it's traditionally been quite hard to go and check when people just make stuff up about the far side of the world.

That's something that was exploited by possibly the greatest con artist of all time—a man who scammed a country by inventing a whole other country. He's just one of the small-time crooks and big-time fantasists we'll meet in the next chapter, "The Scam Manifesto," which explores our eternal fascination with grifters. From the bafflingly simple scam of the original confidence man, William Thompson, via the Soviet grifter who played bureaucracy at its own game, to the Frenchwoman who lived the high life for decades based on the unknown contents of a mysterious safe, we'll look at history's most incredible charlatans and ask the question: How much was a con, and how much did they believe themselves?

If there's one thing everybody knows about politicians, it's that they lie. The leaders of our great nations are not always honest with us. Now, in fairness to (some) politicians, that might be a little unfair—but the untruths of statecraft still deserve their own chapter. In "Lying in State," we'll examine the ignoble arts of political deception: from spin to conspiracy theories to failed cover-ups to wartime propaganda.

Wherever there's money to be made, there'll be someone willing to twist the truth to make it. In "Funny Business," we'll look at two of the biggest culprits: the worlds of commerce and medicine. Business has rested on deceptions small and large throughout history, from Ea-Nasir, the ancient Mesopotamian copper merchant who took people's money but never produced the copper (prompting history's first recorded customer complaint letters), to Whitaker Wright, who made himself a fortune in the nineteenth century on a series of frauds. And we'll meet a selection of history's snake-oil salesmen, from the infamous "goat-gland doctor"—a new-media pioneer with political ambitions who got rich from surgically implanting goat testicles into impotent men—to the man who sold a few hours in his high-tech sex bed to the great and good of London for vast sums of money.

By this point, we'll have met many of history's most impressive liars. But if we think that liars are the only problem we have, then we're in for a nasty shock. It turns out that, when humans get together, we're very good at creating myths out of thin air. In "Ordinary Popular Delusions," we'll see how manias, moral panics and mass hysteria lead us to believe some ridiculous things—from the phantom airships that haunted Britain, to the remarkably common belief that something's trying to steal men's penises, and from monster hunts in the American pines to...well, literal witch hunts. When it comes to living lives of truth, it turns out, we're our own worst enemy.

And in the final chapter, "Toward a Truthier Future," we'll ask: What can we do about all this? If lies and bullshit have been ever present throughout history, what does that mean for the knowledge industry—things like science and history and all our other ways of trying to establish facts about the world? Are we doomed to live out our lives in a fog of misinformation, or are there steps we can all take to move the dial back a little toward honesty?

This book will take you on a whistle-stop tour of just a few of history's most incredible lies, most outrageous bullshit and most enduring falsehoods. A lot of what you'll find in here is unbelievable—and yet all of it was believed by somebody. By the end of it, you'll understand why there's never been a Truth Age, and you'll have a newfound appreciation of the wonderful variety of nonsense we've come up with as a species. Bluntly, this book will make you a better, smarter and more attractive person.

Honest. Would I lie to you?

1

THE ORIGIN OF THE SPECIOUS

This is a book about truth— or, more specifically, about things that aren't the truth.

Unfortunately, this means that, before we get any further into the book, we need to have a bit of a think about what "truth" actually is. And, more important, what it isn't.

The thing is, this all gets messy remarkably quickly, because of the sheer variety of ways there are of being wrong. This might come as a surprise to some people. A lot of us assume that there's simply true and false, and moreover that they're easy to tell apart. Unfortunately, it's not quite that simple. Throughout history, those who've pondered the nature of truth and its opposites have realized one central principle over and over again: while there's an extremely limited number of ways of being right, there's an almost infinite number of ways to be wrong.

"Truth had ever one father, but lies are a thousand men's bastards, and are begotten everywhere,"[4] the Elizabethan writer Thomas Dekker bemoaned in 1606. Or as the sixteenth-century philosopher Michel de Montaigne put it in his essay "Of Liars": "If falsehood had, like truth, but one face only, we should be upon better terms…but the reverse of truth has a hundred thousand forms, and a field indefinite, without bound or limit."[5]

This book is an attempt to catalogue just a few of those hundred thousand forms.

Ours is far from the first period in history to have become obsessed with truth and the lack of it. Indeed, there's a whole couple of centuries that, in Europe, are sometimes known as

Niccolò Machiavelli: he knew.

the "Age of Dissimulation" because lying was so prevalent—the continent was being torn apart by religious strife from the 1500s onward, and everybody had to wear a mask of deception just to survive. Machiavelli, a man so connected with the art of political deception that we still (rather unfairly) use his name to describe it, wrote, in 1521, that "for a long time I have not said what I believed, nor do I even believe what I say, and if indeed I do happen to tell the truth, I hide it among so many lies that it is hard to find."[6] Let's be honest—we've all had days at work like that.

So concerned with falsehood were people throughout history that they came up with a remarkable variety of ways to identify liars. The Vedas of ancient India proposed a method based on body language, saying that a liar "does not answer questions, or they are evasive answers; he speaks nonsense, rubs the great toe along the ground, and shivers; his face is discolored; he rubs the roots of the hair with his fingers; and he tries by every means to leave the house."[7] Also in India, a few centuries later, was a weight-based method: the accused liar would be put on a set of scales with a perfect counterbalance. They would then get off, the scales would be given a short speech exhorting them to reveal the truth, and the person would get back on. If they were lighter than before, they were not guilty; if they were the same weight or heavier, they were guilty.[8]

(Interestingly this implies a completely different relationship between weight and truth from many occult trials in Europe: in India, lightness was associated with innocence, whereas in Europe appearing unexpectedly buoyant could be enough to condemn someone accused of witchcraft. As such, the Indian approach is a judicial process that makes the rare argument for the benefits of weeing yourself while in court.)

Of course, other cultures preferred simpler, more direct methods of identifying liars, such as red-hot pokers or boiling water. It is unclear if these were any more effective.

For a long time, people have devoted considerable effort to trying to classify the different types of falsehood. It was kind of the theological equivalent of writing a BuzzFeed list. As early as AD 395, Saint Augustine came roaring out of the gate by identifying eight types of lie, in descending order of badness: lies in religious teaching; lies that harm others and help no one; lies that harm others and help someone; lies told for the pleasure of lying; lies told to "please others in smooth discourse"; lies that harm no one and that help someone materially; lies that harm no one and that help someone spiritually; and lies that harm no one and that protect someone from "bodily defilement." (I think, by the last one, he means "cockblocking," but I'm not 100 percent sure.)

These days, of course, we classify lies differently. But, even then, there are subtleties that you might not be aware of. Everybody's heard of white lies—harmless social fictions intended to enable us to all get along without killing each other—but did you know there are other colors of lie? "Yellow lies" are those told out of embarrassment, shame or cowardice, to cover up a failing: "My laptop crashed and deleted that report I said I'd definitely have finished by today." "Blue lies" are the opposite, lies downplaying your achievements, told from modesty ("oh, the report's nothing special; Cathy wrote most of it, really"). "Red lies" might be the most interesting of all—they're lies that are told without any intent to deceive. The speaker knows they are lying, the speaker's audience knows that they're lying and the speaker knows that the audience knows. The point, here, isn't to mislead anybody—it's to signal something to the audience that can't be spoken out loud (whether that's basically "fuck you" or the more benign "shall we all just pretend that didn't happen"). Imagine a couple denying to their neighbors that they had a huge row last night when they know everybody could hear it, and you're in the right territory.

It's often said that a lie can travel halfway around the world while the truth is still getting its boots on. (The question of exactly *who* said this is a thornier matter. It's often attributed to Mark Twain, or to Winston Churchill, or to Thomas Jefferson or to any number of the other usual suspects for quote attribution. These attributions are, of course, all lies. The earliest formulation of it may in fact have been from the iconic Irish satirist Jonathan Swift, who wrote in 1710 that "Falsehood flies, and the Truth comes limping after it.")

Whoever said it, it's certainly true that bullshit can move with remarkable and terrifying speed, as you'll know if you've ever tried to debunk rumors on the internet—that is, in fact, my day job, so, believe me, I get it.

Jonathan Swift, pondering some bullshit.

But, in reality, the reason that untruth so often has the advantage over truth has less to do with fact and fiction's relative speed, or even with truth's impractical footwear choices, and more to do with the sheer scale and variety of falsehoods on offer. For every lie that travels halfway around the world, there may well be thousands that never make it out of the front door. But the sheer number of possible lies out there—unconstrained as they are by the need to match up to reality—provides a huge Darwinian testing ground to find the most compelling and long lasting among them—those zombie untruths that will keep coming back again and again. It's like those species of fish that lay two million eggs, just so that two of their offspring will survive.

The truth, by contrast…well, it's kind of boring. It just sort of sits there, a small gray blob of indeterminate size, familiar yet inscrutable. In addition to being slightly dull, it's also remarkably frustrating; as anybody whose job it is to try to pin down small fragments of truth will testify, it has a nasty habit of slipping out from under your grasp just when you think you've got hold of it.

There are, of course, certain things that are simply, uncontestably true: fire is hotter than ice; the speed of light in a vacuum is a constant; the best song ever recorded is "Dancing on My Own" by Robyn. But once you go beyond these immutable laws of nature, everything gets murky alarmingly quickly. You find yourself saying things like "The best available evidence suggests…" and "Yes, but what about the big picture?" rather a lot. Anyone who has spent time in pursuit of accuracy and evidence understands how every new fragment of knowledge has a tendency only to raise ten more questions; every time you think you're approaching enlightenment, reality recedes further toward the horizon, while you're left drowning in a sea of caveats. Truth, by this measure, is not so much a *thing*; it's more of a long, irritating journey toward a destination you'll never reach.

The myriad untruths our world offers us, meanwhile, are seductive, adaptable and—if we're honest—often tremendous fun.

It's that sheer variety of untruth that this book will look at, because lies are in fact only one manifestation of the "hundred thousand forms" that the reverse of truth can take.

For example, there's spin, the art of political deception. The cunning thing about spin is that it doesn't even necessarily need to lie in order to be dishonest. While many politicians do lie (shock news, I know), the real peak of the spinner's craft is managing to suggest something wholly untrue by saying only things that are true—building a house of nonsense from honest bricks. Then there's delusion, the consistent ability people have to be wrong while convincing themselves that they're right, from the ways we overestimate our own qualities to the ways we can succumb to mass hysteria and mob rule. And then, perhaps the most widespread and damaging of all, there's bullshit.

We have the philosopher Harry G. Frankfurt to thank for our understanding of bullshit—he was the first to devote serious time to analyzing this complex subject, in his seminal work *On Bullshit*. (Yes, Harry Frankfurt is clearly having a great time being a philosopher.)

Frankfurt's key insight is that—despite what you might think—lying and bullshitting aren't actually the same thing. He writes: "It is impossible for someone to lie unless he thinks he knows the truth. Producing bullshit requires no such conviction."

In other words, a liar cares deeply about truth for the same reason that a sailor cares deeply about icebergs. They need to know exactly where the truth resides so that they can take precise and deliberate actions to avoid it. For the bullshitter, by contrast, truth is irrelevant; they are happy to take it or leave it. When bullshitting, a little accidental accuracy may be best regarded as an optional extra; if the bullshit world you are creating sometimes overlaps with the real world, it does you little damage and may even be a helpful bonus. For a liar, on the other hand, the careless admission of an inconvenient fact may prove fatal.

Bullshit operates on dream logic, merrily plowing through inconsistencies because, well, it makes sense at the time. Frankfurt notes that this "indifference to how things really are" is, to his mind, "of the essence of bullshit."

Their effect on the world, as a result, is profoundly different. Lying is a scalpel; bullshit is a bulldozer. If you've looked around at the world in recent times and wondered how these lying liars can get away with their lies so brazenly, and why people won't call out their lies for being lies, well…there's your answer. You were accusing them of the wrong thing. Lying—a tricky, detail-focused and analytical profession—is not necessarily our main problem. Our main problem is bullshit.

And then, beyond all of these varieties of untruth, there's just plain old being wrong.

As I've mentioned, my day job is at a fact-checking organization, and as such we come into pretty regular contact with the whole pantheon of ways that people can be wrong. So much so that, last year, we invented a sort of thought experiment to try to get people to ponder all the different types of errors that might be found in the wild. The idea is to strip away all the confusing, messy stuff that surrounds most things in the world and take each story down to a single, simple factual claim from a single source—one where you can't rely on any other evidence beyond that claim to back up or disprove it. We call it "the clock game," and here it is:

You are startled awake by the insistent ringing of a phone. You open your eyes. You are in an unfamiliar room, dimly lit by a faint light seeping around what you assume is the bathroom door. From the universal design cues that say "sort of but not really like a home," you sense that you're in a hotel room of some kind. You aren't sure where you are or how you got here—but, from the foggy state of your brain, you begin to realize that you're extremely jet-lagged.

You have no idea how long you've been asleep.

You look around the room for some kind of clue. There is no clock visible, and blackout curtains cover the windows, offering no hint of whether it's day or night outside. The bedside phone is still ringing, far too loudly for your comfort. Fumblingly, you pick it up.

"Hey, you made it!" says a slightly too cheery voice at the other end. The voice has an indeterminate accent you can't quite place.

"Buh?" you reply. "Who is this?"

"It's Barry!" says the voice. "Glad to finally connect with you!" You're not sure you know a Barry, but you decide to go with it.

"I, er…" you start, before realizing you don't have anywhere else to go with that sentence. "Uh…what time is it?" you settle on, limply.

"Wait a minute," says this person who claims to be Barry, "let me go and look at the clock."

You hear the noise of a phone being put down and steps receding into the distance. A period of time passes, which could be a few seconds or could be several minutes, you're not sure. The steps return.

"It's five o'clock, mate," says the self-professed Barry.

"Okay," you say.

The point of the game is this: Can you list all the different ways in which your belief about the time could, in this moment, be wrong? Spoiler alert: there are probably more ways than you think! To date, we've got somewhere over twenty, and, even then, we've almost certainly missed a few.

Go on, take a moment and see how many you come up with. Imagine some easy-listening music is playing at this point.

["TAKE FIVE" BY DAVE BRUBECK PLAYS WHILE YOU THINK DEEPLY ABOUT CLOCKS AND ALSO POSSIBLY WONDER IF THE AUTHOR HAS GONE MAD.]

Okay, are you back? Good! Let's take the obvious ones first. Barry's clock might be wrong: it might run too fast or too slow, or it might have stopped completely; or it might run at the perfect speed but have been set to the wrong time in the first place. It might be a really hard-to-read clock, one of those over-fancy designer jobs made out of reclaimed driftwood and glass globes, which looks very nice on the wall but isn't very useful for telling any time other than splinter-past-bauble. It might not be a clock at all. Maybe it's just a painting of a clock. Maybe Barry doesn't have a clock and just got someone to write the time down on a piece of paper earlier that day.

Maybe you and Barry are in different time zones—so, while he was perfectly correct about the time, it isn't true for you. Maybe he rounded it to the nearest hour, for convenience, but that's not actually very useful for you, because you wanted to know if it was closer to half past. Maybe it *was* five o'clock when he looked at the clock, but by the time he got back to the phone, it wasn't anymore.

Maybe Barry was deliberately lying to you, for whichever of the many nefarious purposes that Barries have. Maybe he wasn't lying but was bullshitting, because he can't read the time but didn't want to admit that. Maybe he *thinks* he can read the time but actually doesn't know how clocks work. Maybe he meant to say "nine o'clock" but misspoke.

Or maybe he *did* say "nine o'clock," and you just misheard him. Maybe you're the one who doesn't actually understand how time works, and right now you're thinking, *Ah, five o'clock, so it's almost midnight.* Maybe you supposed that he wouldn't factor in the time it took to get back to the phone, so you assume it's actually something like five past five now, but in fact he'd already done that and so you've overcorrected.

Maybe, in your slightly paranoid state, you *assume* that Barry is lying to you—so now the one thing you think you know for sure is that it definitely isn't five o'clock. But you're wrong. Barry is a good man, and your friend, and he would never lie to you. It really is five o'clock, and your lack of trust has led you astray.

Maybe you and Barry don't even use the same time system.

Maybe he's a NASA engineer working on a Mars project, and his clock is set to the Martian day, which is thirty-seven minutes longer than Earth's.

Maybe "It's five o'clock, mate" wasn't even an attempt to tell you the time, but a code-word check-in for the secret agency you both work for, which you have forgotten all about due to traumatic amnesia.

Maybe time, that mysterious river down which we all must be carried, cannot truly be measured by humans, and all our efforts to do so are but crude approximations.

Or maybe…maybe he just meant a.m. when you assumed he meant p.m.

Now, this may all strike you as, frankly, nonsense—but, in fact, every one of those ways you could be wrong about the time matches up to a real-life example of how bad information gets out into the world. Yes, even the stupid ones, like Barry working on Martian time or him trying to give you a super-spy code word.

Some of the real-world equivalents are pretty obvious; rounding too far, not adjusting for errors (like the time between clock and phone) or not realizing that your source is simply unreliable (like the clock being slow) are all common problems, especially when you're dealing with facts based on data. Trying to tell the time from a stopped clock or a piece of paper matches up with the human habit of being extremely certain about things when it should be clear that we don't actually have any useful information to go on. Barry's Martian clock is a surprisingly common one—people just don't realize that they're using completely different definitions of the same basic concept (remember that Christopher Columbus only "discovered" America because he had the wrong idea about how far away Asia was, which was because he'd worked out the circumference of the earth using a source that he assumed was given in Roman miles but was actually talking about Arabic miles, which are a totally different length).

Weirdly, when researching this book, I discovered that we weren't the first to hit on this kind of thought experiment. In 1936, Vilhjalmur Stefansson, a man with a somewhat checkered career as an intrepid Arctic explorer, took a bit of a career turn and wrote a book called *Adventures in Error*—the quote at the beginning of this book comes from it. In it, he gives a very similar example, only he uses a cow instead of a clock:

> *Take an example: A man comes from out-of-doors with the report that there is a red cow in the front yard...we are confronted with numerous other sources of error. The observer may have confused the sex of the animal. Perhaps it was an ox. Or if not the sex, the age may have been misjudged, and it may have been a heifer. The man may have been color-blind, and the cow (wholly apart from the philosophical aspect) may not have been red. And even if it was a red cow, the dog may have seen her the instant our observer turned his back, and by the time he told us she was in the front yard, she may in reality have been vanishing in a cloud of dust down the road.*[9]

I hope that all this waffle about cows and clocks has convinced you that, if it sometimes seems like we're drowning in a sea of falsehood, there's a very good reason for that: it just has a natural advantage over truth, because there can be so much more of it. But that's not the only advantage it has. There are lots of things about our brains and our societies that allow falsehood to flourish.

For many centuries, we've believed that lying was a uniquely human trait, our original sin. But it turns out that humans aren't the only creatures that lie. For starters, the lives of many animals and plants are founded on deception—think of the opossum pretending to be dead, or the cuckoo sitting parasitically in another bird's nest or the orchid that looks like a sexy lady bee to fool horny male bees into pollinating it. But, you might reasonably say, that's not *lying* exactly—it's just the involuntary end product of many generations of an evolutionary arms race. Which is fair enough, but there's plenty of evidence that some of the smarter animals are perfectly capable of intentional and deliberate deception.

To take one particularly memorable example: in his work "Can Animals Lie?," the semiotician Thomas A. Sebeok mentions a "handsome tiger" living in the Zurich zoo who had learned to deliberately lure visitors toward the bars of his cage by means of "a certain sequence of interesting activities."[10] When the enthralled tourist got close enough, the tiger would—and there's no way to say this delicately—drench them with a powerful stream of piss. The tiger was apparently so pleased with this trick that the zoo management eventually had to put up a sign warning visitors that the tiger was not to be trusted.

That pissy tiger is far from alone. A dolphin at a research facility in Mississippi who had been trained to help clear rubbish from its pool by being given rewards of fish learned to hide pieces of garbage under a rock, which it would then bring to the surface to scam fish on demand.[11] Chimpanzees have been recorded performing a wide range of deceptions. They grin involuntarily when they're nervous—one chimp, being threatened by another chimp behind them, was seen physically pushing its lips back down over its teeth before turning around and bluffing that they weren't scared. Another young male, the least dominant of its group, was seen trying to surreptitiously seduce a lady chimp that the dominant males wouldn't normally let him approach. When one of the older males interrupted him, he covered up his erection with his hands, like a character in a 1970s British sex comedy.[12]

Trickery is built into much of the natural world, so maybe we shouldn't be too hard on ourselves for telling the occasional fib.

It's not simply that deception is natural—it's that it appears to have evolved. One scientific study showed that across all the primates there was a close correlation between the size of the neocortex (the part of the mammalian brain that deals with complex tasks, like language) and the frequency of deception in those species.[13] In other words, a bigger brain equals more lying. The challenges of living in complicated social groups— including the need to sometimes deceive your peers—may well have driven the increasing complexity and size of our brains.

That link between cognitive power and deception is replicated as we grow up. Children generally start telling their first lies at about the age of two and a half, not long after they've started talking. Initially, the first lies are simple "wish fulfillment" lies: "I would like to not be the person who ate the cookies."[14] But, as kids' mental abilities develop, as they get a theory of mind and begin to understand the complex nature of their interactions with others, their skill at lying marches in lockstep.

How deeply embedded is falsehood in our daily lives? Possibly more than you think. Psychological studies suggest that within the first ten minutes of conversation when you meet someone new you will, on average, have told three lies.[15] Other studies suggest that, on average, each of us lies at least once every single day—although those studies are based on asking people to report how often they lie and so are vulnerable to the possibility that the participants are…lying about that.

That's not the only potential issue with asking people how often they lie. In writing this book, one of my original plans was to keep a "lie diary"—to spend several weeks diligently noting and recording every single time I uttered a falsehood. It was going to be an attempt to gain an insight into just how much untruth permeates our lives, even (or especially) for those of us who believe ourselves to be fundamentally honest people. I was excited by this prospect, although also nervous; exactly how many friendships, I wondered, would be destroyed forever by the publication of this book?

In the end, I needn't have worried. Not because it turned

out that I am a beacon of purity and truth (I mean, obviously I *am*), but because every attempt I made to try to record my lies ground to a crashing halt after about a day.

Quite simply, I wasn't able to spot when I was bullshitting.

The thing is, I know for a fact that I *did* tell lies during this time. None of them were especially heinous; I was not doing any massive crimes during the writing of this book. Broadly, they fell into three categories: lies about what I'd already done, lies about what I was able to do in the near future and lies about my social life.

The first category mostly consisted of texts and emails to my publisher and agent insisting that the book was going really well and I'd got a lot written. (Sorry.) The second was primarily to colleagues, asserting confidently that I would get around to that thing I'd promised them next and that I'd definitely have something for them by tomorrow. (Sorry again.) The third was that broad category of little white lies that keep society from falling into a death spiral of mutual recrimination: fabricated excuses for not being able to attend a party; transparently false claims about having only just seen a text; hollow assurances that, yes, you are unquestionably being the reasonable one in this argument and the other person sounds like a complete asshole who definitely doesn't maybe have a point.

(That last category would probably have been considerably larger if it wasn't for the fact that I was, well, trying to write a book at the time and so spent many months turning down invitations to go to the pub for perfectly genuine reasons—namely, that I had to concentrate on the important business of staring blankly at a screen and not writing anything. Pro tip, introverts: an imminent book deadline is an excellent and entirely genuine excuse for getting out of social engagements!)

Most of the time I was well aware that these were lies at the time I told them, with the occasional exception of promises that I'd get something done (which were sometimes based on pure and simple delusion that I could work solidly for every one of the thirty-six hours that I understand there to be in a day). And yet something happened in my brain during the act of telling them—a switch would flip, and I would temporarily blank out that I was dishing up small servings of horseshit. It's something I'd never really noticed until I set myself the task of noting down all my little white lies: I simply wasn't able to recognize them in the moment. It was like my brain had a self-defense mechanism against telling on itself.

I have no idea if anybody else's brain works like this. It's entirely possible that I might have just inadvertently discovered I'm a psychopath. But, at a guess, I'd say the chances are pretty good that this happens to quite a lot of people.

Liars lie; bullshitters bullshit. That much is easy. But what's really interesting isn't why people say things that aren't true— that's always going to happen. No, the really interesting question is why some lies stick around—why, despite all our professed reverence for truth and all the structures we've set up as a society to identify and root out falsehood, some untruths become widely believed. In other words, how do bullshitters get away with it?

The reason is that, in addition to their numerical advantage over truth, there are some structural reasons that mean falsehoods have the upper hand. Throughout this book, we'll keep encountering the seven main ways that untruths spread and take hold.

The Effort Barrier

You get an effort barrier when relative difficulty of checking the truth of something outweighs its apparent importance. The key thing about this is that it works at both ends of the scale: it applies to things that would be relatively easy to check but are so trivial that nobody bothers, and to things that are clearly pretty important but are also really hard to check. The reason that sixteenth-century explorers could get away with claiming that a race of twelve-foot-tall giants lived in Patagonia is the same as the reason you can usually get away with upgrading your AP math grade from a B to an A on your CV. Yes, someone *could* check, but are they really going to bother?

This is something that seasoned bullshitters understand instinctively. It's simply inefficient to craft untruths that are built to withstand far more scrutiny than they'll ever receive. A talented liar pitches their falsehoods, both big and small, just on the far side of the effort barrier.

Information Vacuums

We often like to think of Truth and Lies as being in some kind of eternal battle. But one effect of the effort barrier is that, quite a lot of the time, Truth never even shows up to the fight. There are an awful lot of things in the world that we just don't really know anything about. And, in the absence of information, we tend to lower our guard whenever something that *claims* to be information shows up—even if there's no good reason to believe it.

This all ties in with the cognitive bias known as "anchoring": our brain's tendency to latch on to the first piece of information we get about any subject and give it far more weight than anything else. When there isn't good information on something, crappy information will always flood in to fill the void—and, a lot of the time, it refuses to budge, even when better information finally shows up.

The Bullshit Feedback Loop

None of us can work out the entire world by ourselves. All of us have to rely on others for our information. This is a good thing—together, we can find out much more about the world than we ever could alone—but it does come with some downsides. And one major downside is the bullshit feedback loop. You get one of these when a dodgy piece of information is repeated, but, rather than the repetition being seen for what it is (just somebody copying someone else, adding no extra level of verification to the claim), it instead gets treated as confirmation that the original dodgy info was accurate. If this goes on too long, the problem expands. It's no longer merely that the claim gets repeated; eventually, it becomes so established that people start adjusting what they say to accommodate the dodgy facts—*everybody* knows it's true, so even if you're staring directly at evidence proving it to be false, it's probably because there's something wrong with your eyes.

So person A tells person B something wrong, and then tells person C too. Person C is skeptical, but then person B tells it to them as well, and person C interprets that as a second source and is now convinced. Person C runs to tell person D the exciting news, whereupon person D tells person A, who takes it as evidence that they were right all along. Meanwhile, people E, F, G, H and I have also heard the same thing from multiple people, and it's become accepted as common knowledge. At this point, person J tentatively asks, "Are we actually sure about that?" and is promptly burned as a heretic by the rest of the alphabet.

Or, to take a familiar example, it's the thing where a newspaper copies a fact from Wikipedia and then gets cited on Wikipedia as evidence that the fact was correct.

Wanting It to Be True

There are a whole host of things our brains do that make us particularly bad at sniffing out the difference between truth and not-truth. These have a bunch of technical names you've probably heard of—things like "motivated reasoning" and "confirmation bias"—but they all basically come down to the fact that, when we *want* to believe something, working out whether or not it's actually true comes pretty far down our brain's list of priorities. It doesn't really matter whether it's something that supports our political stance, something that matches up with our prejudices, or just basic wish fulfillment of the "maybe I *have* won the lottery in Spain, despite never entering" type, we'll cheerfully come up with spurious reasons to assign even the most ridiculous claim credibility, cherry-picking only the evidence that supports it, while blithely ignoring that vast mountain of evidence that says it's crap.

The Ego Trap

Even when falsehoods do get unmasked, there's something that often stands in the way of the truth spreading as easily as the lies that got their boots on first: simply, we really do not like to admit that we're wrong. Our brains don't like doing it, and there are a whole host of cognitive biases that push us away from even acknowledging that we might have fucked up. And, if we do come to realize that we've been taken in by something false, there are a multitude of social pressures that make us want to cover it up. Once bullshit has us in its grasp, we can be rather unwilling to break free.

Just Not Caring

Even when there is an opportunity to push back against untruth, we don't always take it. We might think it's just not important whether something's true or false (especially if we like the lie). But, equally, we might think that pushing back would be ineffective and so not bother. We might believe that lying is so widespread that we get overwhelmed by the scale of it all and just give up. Equally, we might think that…well, if everybody's doing it, I should get in on the game too.

All of these are understandable, but bad.

Lack of Imagination

Perhaps one of the strongest advantages that untruth has is, quite simply, we don't understand all the myriad and surprising ways that it can manifest. This makes sense—after all, we have to live our lives on the assumption that most of the stuff we're told is true; otherwise we'd descend into a spiral of gibbering paranoia. But this can lead to us radically underestimating the likelihood that something might not be true. We assume that if we read something in the news, then it's probably true. We think that if someone seems trustworthy, then they aren't trying to scam us. We believe that if lots of eyewitnesses said they saw something, then there must have been something there. None of those assumptions is as reliable as we might think.

Fundamentally, we just haven't been paying enough attention to the business of falsehood. We haven't studied it and we don't talk about it, with the result that we don't always recognize it when we see it.

Hopefully, by the end of this book, that won't be a problem anymore.

2

OLD FAKE NEWS

Titan Leeds was dead. There was no doubt whatsoever about that.

A hardworking and honest man—he had been a successful publisher in the city of Burlington, New Jersey, prior to his sad demise—Mr. Leeds passed away on Wednesday, October 17, in the year of our Lord 1733, at around half past three in the afternoon. The unhappy news of Leeds's death was somberly recorded, printed and distributed in black-and-white for all to read: "'tis undoubtedly true that he is really defunct and dead,"[16] the story of his passing reads, in part. And, even though it had been predicted in advance that he was not long for this world, the news of his loss at a relatively young age—he was still in his early thirties—must nevertheless have come as a shock to many in Burlington, a bustling community beside the Delaware River that had grown rapidly since its founding by a group of Quakers some five decades previous.

The person who was most shocked by it would probably have been Titan Leeds himself, as he was pretty darn sure he was still alive.

We can only guess at his *exact* reaction. But there's good reason to believe that the very much non-deceased Mr. Leeds must have been, shall we say, quite put out upon reading the news of his untimely death. I mean…that's the sort of thing that would throw you a bit, right? Any of us could be forgiven for freaking out a little. But it must have been especially confusing in the world of the 1730s, because, at that time, Leeds wouldn't really have had many reference points for what was happening to him.

In our age, the unsettling experience of reading about your own death is, thankfully, still a pretty rare one—but we are at least broadly *aware* that it's a possibility. We've probably all heard tales in the news of people it's happened to: corpses mistakenly identified, or obituaries accidentally published early. "Reports of my death have been greatly exaggerated" is such a well-known quote that it's now virtually a cliché (even if a pedant would point out that Mark Twain never actually said *quite* those words).[17] In 1980, the *New York Times* ran an obituary of the notorious hoaxer Alan Abel[18]—an editorial decision with seemingly obvious pitfalls, as became very apparent the following day when Abel held a press conference to announce that he'd staged his own death in order "to gain publicity."[19] (When Abel finally passed away, a full thirty-eight years later, the *New York Times*'s second go at his obituary wryly recorded that he "apparently actually did die."[20])

In other words, for us, premature death notices are "a thing." Not only do we know that this kind of stuff sometimes happens, but many of us have probably, at some point, imagined what it might be like if it happened to us. (Admit it: "People mistakenly believe I'm dead, so I get to find out what everybody *really* thinks of me" is a thought that's occurred to you in your darker moments.) In 2009, when the combination of a spoof website and the always overactive Twitter rumor mill temporarily killed Jeff Goldblum, the actor ended up going on *The Colbert Report* to deliver his own eulogy, which I think we can all agree is the classy way to handle something like that.[21]

But, for Titan Leeds—living, as he was, far nearer to the dawn of the era of mass media—this must have all been strange and new. Having people think you were dead because of something they'd read must have been far weirder even than it would be for us...and also far more infuriating. Not least because his own efforts to debunk the news were, shall we say, not 100 percent successful. Despite Leeds angrily insisting in print that he was very much alive, reports continued to appear for several years afterward, confidently asserting that he was definitely super dead. To add insult to injury, these reports insisted that whichever imposter was now writing angry screeds about still being alive under the name of the late Mr. Leeds should stop at once, because it was besmirching the beloved memory of the departed.

This all happened because the story of Leeds's death hadn't simply been an innocent mistake—a clerical error, say, or an unfounded rumor credulously repeated. It was actually a deliberate and gleeful falsehood, spread for two classic reasons: profit and mischief. It was a no-holds-barred (and remarkably successful) effort to boost sales on the part of an upstart publishing rival. It was Titan's extra misfortune that this rival happened to have a particularly impish sense of humor. And, if Leeds was annoyed about that, he'd probably have been absolutely *furious* to learn that the two-bit huckster who'd fabricated his death as a cheap marketing stunt would, in the following decades, go on to become the most celebrated intellectual hero of the nascent United States.

In short, Titan Leeds was having a fairly brutal early encounter with what can only be described as "fake news."

Titan's post-truth predicament started for the simple reason that a rival almanac maker had turned up just down the river. In the America of the 1730s, almanacs were big business, and Titan Leeds was at the top of the game. He'd inherited publishing duties on the *Leeds Almanac* from his father, Daniel, when the old man was forced into retirement. Daniel Leeds had been born into a Quaker family, originally from Leeds in England (the small one in Kent, not the big one in Yorkshire).[22] In the face of rising persecution, the Leeds family immigrated to the Americas in 1677, in an effort to escape the religious strictures of the Old World—only for Daniel to run smack-bang up against the religious strictures of the New World.

A thoughtful and self-taught man, much given, in his youth, to spiritual visons and occasional bursts of weeping, Daniel Leeds had a very particular and somewhat unorthodox personal philosophy—one that combined heterodox Christian mysticism with a deep love of science. It was his desire to spread the truth as he saw it that brought him to publishing—first with a pioneering almanac, and then with a grand philosophical and theological tract that represented the apex of his life's work. He was utterly crushed when, angered by his nonconformist ideas and heavy use of astrology, his fellow Quakers in the community that had founded Burlington rejected his works, suppressing his first almanac and destroying almost every copy of his book.

But Daniel Leeds was unbowed, and, rather than settle for the quiet life, he returned to producing his almanac with renewed vigor—in addition to engaging in a long-running and incredibly bitter pamphlet war with his neighbors, a rolling series of feuds that ended up with Leeds being accused by one nemesis (he had several) of being a literal devil. He was, they wrote, "Satan's harbinger." Eighteenth-century Quaker beef could be savage.

Such notoriety may have made for some awkward moments on the streets of Burlington, but it wasn't necessarily bad for business, and, from its origins as one of the first true almanacs

in the American colonies, the *Leeds Almanac* found a sizable audience. By the time the acrimonious fallout from an ill-advised
political alliance finally pushed Daniel to hand over the reins of
the almanac to his teenage son, Titan, in 1714, it had a decades-
long reputation as the leading almanac in the area.

The trouble with being a market leader, of course, is that it
paints a big old target on your back for any rivals trying to enter
the market. Which is exactly what an ambitious young chap
named Benjamin Franklin took aim at when he decided to get
into the almanac game.

These days, Franklin is remembered as one of the core Founding Fathers of the USA, the man who, above all others, was the
great intellectual heavyweight of the American independence
movement. Franklin was a multitalented genius, whose legacy
stretches from pioneering experiments with electricity to creating America's first public lending library, and from establish-
ing the US postal system to the invention of bifocal glasses. I
promise that I'm not going to make a habit of cut-and-pasting
from Wikipedia in this book, but, to give you a sense of just
how much of an irritating overachiever Benjamin was, the opening lines of his page identify him as "a leading author, printer,
political theorist, politician, Freemason, postmaster, scientist,
inventor, humorist, civic activist, statesman, and diplomat."[23]

Honestly, it's exhausting just reading it. *Stay in your lane, Benjamin.*

But back in 1732, Franklin was still in his twenties and wasn't
yet a leading anything. Having fled his native Boston at the age
of seventeen to escape his elder brother's shadow, he'd recently
set up shop as a printer in Philadelphia, the swiftly growing city
just down the river from Burlington. (These days, thanks to a
few centuries of urban sprawl, Burlington is now a suburb of
Philly). Franklin was good at his job, with a profitable newspaper, the *Pennsylvania Gazette*, already in his burgeoning publishing stable. But if you wanted the big bucks back then, you

really needed your diversified portfolio of media brands to extend into the almanac space.

Benjamin Franklin: annoying overachiever.

Almanacs, in case you're not familiar with them, were effectively a guide to useful information that you might want to know in the coming year. Much as in the following centuries newspapers would do the job of bundling together sports results, TV listings, a bit of opinion, weather forecasts and some light astrology into something that people wanted to buy, almanacs did the same with...well, a bit of opinion, weather forecasts and some light astrology. (The TV listings were slightly less of a big deal in the 1730s.) For communities that still had farming at their hearts, the promise of such knowledge—when the sun would rise and set, when the tides would be high, when the seasons would change—was essential. One major almanac of the time, published by Nathaniel Ames in Massachusetts, had sales upward of fifty thousand copies a year—huge numbers for a still-young publishing industry.[24] You can see why Franklin wanted a piece of that action.

And so it was that, in 1732, he launched *Poor Richard's Almanack*, writing under the pseudonym of "Richard Saunders," whom he characterized as an impoverished stargazer forced into publishing by a demanding wife who insisted that he do something to earn a living. (Franklin really loved a pseudonym. Spoiler warning: this is not the only time a pseudonym of Ben Franklin's will play a significant role in this book.)

By this point, feuds between rival prognosticators were already a well-established facet of the normally mild-mannered American almanac scene, with some competitors fulminating against their rivals in yearly doses of invective. (In 1706, one Boston almanac author, Samuel Clough, instructed his rival, Nathaniel Whittemore, to—and this is an exact quote—"jog on."[25]) But, whereas most of these feuds basically came down to saying, "You're rubbish at doing almanacs," Franklin chose a more sly approach to taking a pop at his main competitor—one that was also a lot funnier. He had "Saunders" write in his introduction that, honestly, he would have jumped at the opportunity to publish a profitable almanac many years earlier, if it wasn't for the fact that, out of the kindness of his heart, he didn't want to ruin the business of his "good Friend and Fellow-Student, Mr. Titan Leeds."

The only reason he'd now changed his mind, he said, was that very sadly this wouldn't be a problem for much longer—because Titan Leeds was going to die shortly. Or, as he put it, "inexorable Death, who was never known to respect Merit, has already prepared the mortal Dart, the fatal Sister has already extended her destroying Shears, and that ingenious Man must soon be taken from us."[26] We pause here briefly to note that *The Fatal Sister Has Extended Her Destroying Shears* is an absolutely brilliant title for a metal album that is currently just sitting there unclaimed.

Franklin predicted that Titan Leeds would die, "by my Calculation made at his Request, on Oct. 17. 1733. 3 ho. 29 m. P.M. at the very instant of the conjunction of the Sun and Mercury," adding for a bit of color that there was a disagreement with Leeds about the exact date: "By his own Calculation he will survive till the 26th of the same Month."

It's a good joke, although it wasn't even one of Franklin's own—he'd pinched it from Jonathan Swift, who, in 1708, had pulled the exact same stunt on an astrologer and almanac author named John Partridge, predicting in a fake almanac published under a pseudonym that "he will infallibly die upon the 29th of March next, about eleven at night, of a raging fever."[27] Franklin, no fan of astrology, was certainly aware of Swift's hoax (and the fact that Daniel Leeds had been an advocate of Partridge) and was winking at any in-the-know readers who would recognize it too.

Unfortunately for Titan Leeds, he was very much *not* in on the joke. Where his father had been blessed with a decent sense of humor, Titan was (in the words of one scholar) "a serious, self-righteous, gullible, practical man who appeared to take things at face value."[28] As a result, he did what any playground bully could tell you was the absolute worst thing to do: he took the bait. Responding to "Poor Richard" a year later, in his almanac for the year 1734 (almanac feuds were a little slower than Twitter), he attacked his rival for "gross Falsehood," branded him "a Fool and a Liar," and proudly boasted that "notwithstanding his false Prediction, I have by the Mercy of God lived to write a diary for the Year 1734, and to publish the Folly and Ignorance of this presumptuous Author."[29]

That was the only excuse Franklin needed, and so he took his hoax to the next level. First in his 1734 almanac, in which he expressed shock at the terribly unkind things written about him and suggested that this indicated his dear old friend Leeds almost certainly was dead, and somebody meaner was writing his almanac in his place. And then, in the following year's edition, he announced that he had now confirmed that Leeds had indeed died on the predicted day, before lamenting that he had "receiv'd much Abuse from the Ghost of Titan Leeds, who pretends to be still living, and to write Almanacks in spight of me and my Predictions."

Exactly how long Franklin would have gone on taunting Leeds in his almanacs, and how Leeds would have reacted to the continued proclamations of his death, is something that we will sadly never know. That's because, in 1738, just when the whole thing was in danger of becoming really confusing, Titan Leeds simplified the situation by actually dying.

That, you would think, should have been the end of it—not least because most people's reaction to the sudden real-life death of the person whose death they'd been joking about would probably be to, I don't know, feel a bit guilty and not mention it again. That would be your reaction, right?

That was not Franklin's reaction.

No, instead, in 1739, he *published a fake letter from the ghost of Titan Leeds*, confirming that "Poor Richard" had been right all along, he really had died in 1733, and asserting that the almanac the real Titan Leeds had been publishing for years was the work of imposters.

Let us speak plainly: Benjamin Franklin was a massive fucking troll.

He was also a successful troll, because it worked. *Poor Richard's Almanack* became a huge hit, while the *Leeds Almanac* went into decline and ceased publication about a decade later. Franklin's almanac was sharper and more entertaining than its competition, and his business practices were more ruthless. He hadn't just been having a cheeky pop at astrology in his writing; he'd also been reminding the audience of the Leeds family's association with strange beliefs and those old slurs about them being "Satan's harbinger." The fact that what he was writing wasn't actually true…didn't seem to matter very much.

Which, in many ways, is the point of this book. Because, as we'll see, throughout history, when we've been faced with the choice between what's true and what's a good story, we tend to go for the good story.

But let's back up a bit: we need to look at exactly how the strange and confusing new world of mass media that Titan Leeds was living in came about. The idea of "news," and the hunger for it, was of course not new. People have always wanted to know stuff, and particularly to know stuff that somebody else doesn't know: what's happening beyond the horizon, or on the other side of a closed door or behind someone's back.

This was particularly the case in earlier centuries when travel was harder and rarer; news moved no faster than a horse, and arrived rarely if ever—and so was hungrily consumed whenever it showed up. In Wales in the eleventh century, the monks at two remote monasteries a hundred miles apart were so eager for fresh information that once every three years they would do swapsies, each sending a news-monk on the hazardous journey across Snowdonia to spend a week living at the other monastery, where they would deliver all the latest gossip.[30]

But from the middle of the fifteenth century onward, things would begin to change dramatically. Disentangling exactly what caused what during this period in Europe's history is…tricky, because bluntly there was a huge amount of stuff going on all over the place: almost constant wars, religious schisms aplenty, the discovery of new lands, more contact with other cultures and the rediscovery of ancient texts from previous ages. But in the interests of simplicity, let's say that there were three major

changes during this time that would transform our basic human
desire to know stuff into something far more consequential: an
information explosion that would shape the world in strange
and profound ways.

One of these was the gradual development of reliable and extensive postal networks. Another was the rise of the merchant
class as international trade expanded, a new wealthy elite with
connections and interests that could span a continent, for whom
the latest news wasn't merely interesting—it was extremely valuable. And the third, of course, was the innovation of the printing press.

The postal networks meant that all of a sudden news didn't
need to be delivered in person: one individual could both send
and receive news from the comfort of their home, without the
need for dangerous travel or an army of messengers (or indeed
an actual army, to make sure the messengers reached their destination safely). The letter quickly became the favored medium
for news addicts of all sorts—and eventually, it became a business opportunity too.

By the late sixteenth century, a new profession had emerged:
the newsletter writer. Starting in the great cities of Italy, these
novellanti would gather all the latest and most reliable information
from their contacts and write them up into manuscript letters
that would be posted across the continent to their subscribers:
wealthy politicians and businessmen who paid a hefty amount
for such a service. Savvy and impeccably well connected, with a
business built on the reliability of their information, these people were, as Andrew Pettegree describes them in his book *The
Invention of News*, "the first news agencies."[31]

Meanwhile, in 1439, Johannes Gutenberg had introduced the
movable-type printing press to Europe, and in doing so set off a
Sudden and Confusing Change Bomb, the shrapnel from which
would fly with reckless abandon across the continent for many
centuries. All of a sudden, the ability to communicate to wide

audiences wasn't limited by how many scribes you could afford; the ability of the establishment powers to act as gatekeepers of information began to erode. From the beginning, print was an extremely capitalist enterprise: mostly done by commercial firms on a for-profit basis, highly competitive and largely un-regulated by the powers of the state or the church (at least, until they worked out that people were doing Protestantism with it). For centuries, the price of a book had remained fairly stable, at "fuck loads." After Gutenberg came along, it dropped at a rate of around 2 percent every year for more than a century. That might not sound like a huge amount on a year-by-year basis, until you think about how that works out in compound terms over many decades: in 1450, a single book would set you back the equivalent of many months' average wages. By 1600, it could cost you less than a day's pay packet.[32]

These sets of developments—business, post and printing—toddled merrily alongside one another for a century and a half, cheerfully causing havoc left and right, until in Strasbourg in 1605 they finally collided thanks to a young man called Johann Carolus.

Carolus was a bookbinder and bookseller by trade—but cru-cially, he'd also recently got himself a side hustle in the newslet-ter game. Strasbourg was a great place to do that from, a buzzing hub for both business and postal networks. But of course, the handwritten newsletter business had a natural ceiling on how much you could grow your trade: namely, how fast you could write. And so it was that Johann looked at the two sides of his income: on the one hand, printed books. On the other, labori-ously hand-crafted letters. Printed books; handwritten letters. Hmm.

Johann Carolus put two and two together, and made THE NEWS.

Essentially, Carolus invested in disruptive new technology in order to scale his media start-up during a growth-hacking phase. The product of that moment of inspiration was the super snappily titled *Relation aller Fürnemmen und gedenckwürdigen Historien*, the world's first-ever print newspaper. It wasn't much like today's newspapers, granted: more like a small book in its quarto format, and lacking modern innovations such as pictures, headlines or being interesting. *Relation* stuck very firmly to the manuscript newsletter format of a dry, undifferentiated list of factual announcements about which important people were currently in which cities, with no effort made to explain who those people were to anybody who wasn't elite enough to know already. (I mean, okay, in that respect it was quite like a modern newspaper's diary column.)

A woodcut of somebody flogging Relation.

But *Relation* was a huge success, and within a few years the idea of the newspaper was being copied in more and more cities across northern Europe. The second newspaper, *Aviso Relation oder Zeitung*, began publishing in Wolfenbüttel in 1609. Frankfurt, Berlin and Hamburg followed up with their own newspapers in the next decade;[33] by 1619, Amsterdam had two competing newspapers.[34] During the seventeenth century, around two hundred newspapers were founded in Germany alone.

The notion was less popular in southern Europe, however; the Italians, pioneers of the manuscript newsletter, turned their noses up at this newfangled nonsense. They weren't the only ones: while the news explosion of the early seventeenth century was greeted with glee by many of the information-hungry population, it also provoked mockery, derision and alarm—much of it in ways that are strikingly familiar to the modern reader.

Anxiety over false news, especially among elites who worry about no longer being gatekeepers; a lack of trust in professional news outlets, contrasted with too much trust in information passed on by someone you know personally; widespread fears about the effects of information overload; disdain for people with "news addiction." They're all prominent features of our twenty-first-century infopanics, but each one of them was also commonplace in the seventeenth century. Often in exactly the same words.

Take news addiction, for a start. Very quickly the Germans came up with a word for this: *Neuigkeitssucht*, which does literally mean "news addiction"[35] and which was despairingly described as the "horrible curiosity of certain people to read and hear new things."[36] In the Netherlands, those obsessed with the latest news were mocked for their addiction: a pamphleteer from the south mocked the northerners because of their insatiable news habit, having them say, "We must read the new Tidings, or we shall not have any patience." The English playwright and satirist Ben Johnson mocked both the production and the consumption of news in several plays in the 1620s, notably *News from the New World Discovered in the Moon* and *The Staple of News*.

Not only was there a great deal of eye-rolling about this insatiable hunger for news, but there was widespread anxiety about the awful effects of this explosion of printed material on both people and society. Just as it is today, information overload was a profound concern, spoken of in apocalyptic terms. In 1685, the French scholar Adrien Baillet wrote apocalyptically: "We have reason to fear that the multitude of books which grows every day in a prodigious fashion will make the following centuries fall into a state as barbarous as that of the centuries that followed the fall of the Roman Empire."[37]

(Also much as it is today, this information overload was talked about as though it was a totally new phenomenon, unique to the age they were living in; in fact people have been moaning about there being too much stuff to read for millennia. It's even in the Bible—"Of making many books there is no end; and much study is a weariness of the flesh," moans Ecclesiastes 12:12. Meanwhile, in the first century, the Roman philosopher Seneca was complaining that "the abundance of books is distraction.")

But while that sense of shame at your unread book pile may be a timeless feeling, for the people living at the beginning of the news age there were good reasons to feel that things were getting a bit much. For starters, there really was a lot going on.

As Robert Burton wrote in his 1621 emo classic *The Anatomy of Melancholy*: "I hear new news every day, and those ordinary rumors of war, plagues, fires, inundations, thefts, murders, massacres, meteors, comets, spectrums, prodigies, apparitions, of towns taken, cities besieged in France, Germany, Turkey, Persia, Poland, &c., daily musters and preparations, and such like, which these tempestuous times afford, battles fought, so many men slain, monomachies, shipwrecks, piracies and sea-fights; peace, leagues, stratagems, and fresh alarms. A vast confusion of vows, wishes, actions, edicts, petitions, lawsuits, pleas, laws, proclamations, complaints, grievances are daily brought to our ears. New books every day, pamphlets, corantoes, stories, whole catalogues of volumes of all sorts, new paradoxes, opinions, schisms, heresies, controversies in philosophy, religion, &c. Now come tidings of weddings, maskings, mummeries, entertainments, jubilees, embassies, tilts and tournaments, trophies, triumphs, revels, sports, plays: then again, as in a new shifted scene, treasons, cheating tricks, robberies, enormous villainies in all kinds, funerals, burials, deaths of princes, new discoveries, expeditions, now comical, then tragical matters. Today we hear of new lords and officers created, tomorrow of some great men deposed, and then again of fresh honours conferred; one is let loose, another imprisoned; one purchaseth, another breaketh: he thrives, his neighbour turns bankrupt; now plenty, then again dearth and famine; one runs, another rides, wrangles, laughs, weeps, &c."[38]

It's possible that Robert Burton may have needed to unplug for a little bit. Take a break to focus on his well-being. Maybe a long weekend away.

Burton doesn't stop there. Like Baillet, he looked at the sudden profusion of print and predicted a coming book apocalypse: "Who could be such a greedy glutton of books, who can read them? As already, we shall have a vast Chaos and confusion of Books, we are oppressed with them, our eyes ache with reading, our fingers with turning."[39]

And elsewhere in *The Anatomy of Melancholy* he appears to be complaining about a profusion of clickbait and hot takes, noting that "it is a kind of policy in these days, to prefix a fantastical title to a book which is to be sold… As Scaliger observes, 'nothing more invites a reader than an argument unlooked for, unthought of, and sells better than a scurrile pamphlet.'"[40] (I don't know if Twitter is looking for a new corporate slogan, but I reckon "an argument unlooked for" would fit pretty well.)

That criticism of "scurrile pamphlets" was also common in

the seventeenth century. While the early newsletters, with their elite subscribers, traded on a reputation for trustworthy and reliable information, the same couldn't be said for everything that was bring printed. While many were addicted to the latest news, there was also widespread distrust of it.[41] People were skeptical of what they read in print: many still believed that manuscript letters were inherently more trustworthy; the most trustworthy information was that delivered in person by someone you knew.[42]

Simply put, lots of people thought there was a lot of fake news around.

They may have had a point. To take just one example: a famous pamphlet that was published in 1614, under the snappy title of *True and Wonderful: A Discourse relating a strange and monstrous serpent (or dragon) lately discovered, and yet living, to the great annoyance and divers slaughters both of men and cattell, by his strong and violent poison, in Sussex two miles from Horsham, in a woode called S. Leonards Forrest, this present month of August, 1614. With the true generation of serpents.*

Now, Horsham may not seem like an especially promising location for dragon hunting: it's a small, pleasant market town, today perhaps best known to many people as the place where you realize you've got on the wrong half of the train and are now going to Bognor Regis. England has many vast, dark and impenetrable ancient woodlands where dragons might plausibly lurk, but Saint Leonard's Forest probably wouldn't be near the top of that list.

But that didn't stop the publisher in question, one John Trundle, putting out the "certain and too true" report of the nine-foot-long dragon, with black-and-red scales, which could run as fast as a man, left a toxic trail behind it like a snail and could spit poison twenty-five yards—by which means it had already killed two people. Most ominously, it had two large budding growths on either side of its body, suggesting that it was in the process of growing wings.[43]

Trundle was exactly the kind of publisher who provoked the widespread skepticism of news in the age: he had a long and infamous history of publishing, basically, trash. If it was implausible but eye-catching, he'd publish it. He was widely attacked by commentators and rivals, as were the many other publishers who traded in the sensationalist and the gory—one anonymous pamphleteer in 1617 blasted the plethora of "fond fables of flying Serpents, or as fond delusions of devouring Dragons, of Men or women burned to death miraculously without fire, of dead men rising out of their graves"[44] that were making it into print. (The pamphlet making this criticism was about "a mighty sea monster, or whale" that had supposedly washed ashore in Essex.)

The anxiety around false news in the seventeenth century was at its strongest in the establishment, who were bluntly not happy about the people being able to just print and distribute anything they wanted. In England this came to a head in the late 1600s, in a country still in turmoil following the English Civil War and the subsequent Restoration. Laws regulating printing presses were introduced, with the king's forces having the power to search premises for illegal presses. It wasn't just printing that they were worried about either—in a classic example of trying to suppress the medium because you're worried about the message, the elite were also freaking out about coffeehouses.

Coffee, much like the newspaper, was a new and frightening phenomenon. London's first coffeehouse was established by a Greek immigrant in 1652, and it quickly became a runaway success. Imitators rapidly sprang up, and within a few decades coffee was established as a vital part of the city's lifeblood. Not only were people drinking coffee, but—to the horror of the establishment—they were having very intense discussions about politics in the coffeehouses. Some of them may even have been spreading fake news while doing so! It couldn't stand.

On December 29, 1675, King Charles II decided that enough was enough and issued "A Proclamation for the Suppression of

Coffee-Houses," in which it was noted that "in such Houses, and by occasion of the meetings of such persons therein, divers False, Malitious and Scandalous Reports are devised and spread abroad, to the Defamation of His Majesties Government, and to the Disturbance of the Peace and Quiet of the Realm."[45] Under the proclamation, all coffeehouses in England (and in Wales, and in the disputed Scottish border territory of Berwick-upon-Tweed) would be forced to close in just twelve days' time, on January 10. The reaction from the caffeine-dependent great and good of London was swift and grumpy: absolutely no god-damn way would they be deprived of their coffee. Charles II was forced to back down and canceled his coffee ban days before it was due to come into effect.

In October 1688, King James II tried again, this time focusing on the message, issuing a proclamation "To Restrain the Spreading of False News." Punishment would be inflicted on any "Spreaders of false News, or Promoters of any malicious Slanders and Calumnies," particularly any "who shall utter or publish any Words or Things to incite and stir up the People to Hatred or Dislike of Our Person, or the Established Government."[46] You can understand why he might have been nervous: at that point a Dutch fleet was preparing to invade England. Unfortunately for James, the attempt to suppress false news didn't help very much: he was deposed by his daughter and fled the country a little over a month later.

England's press-licensing laws were allowed to lapse by 1695, and the result was a second explosion of news outlets. This brought with it all the problems that we still see in the media today. By 1734, the *Craftsman* had already identified one of the key structural problems with the press—namely, their tendency to copy stuff from one another until there's a full-on bullshit feedback loop underway: "When a piece of false intelligence gets into one paper, it commonly runs thro' them all, unless timely contradicted by those who are acquainted with the particular circumstances."[47]

This was exacerbated by the rise of the press across the ocean in America—the flow of information between England and its colony provided many more opportunities to copy news from each other, but with additional effort barrier to actually checking what was really going on across the sea. Wild rumors and complete fabrications about what was happening across the pond would bounce back and forth between England and America, becoming exaggerated with every telling.

Perhaps the best example of this is the way that a courtroom speech by a supposedly hard-done-by woman was sent back and forth across the Atlantic for several decades in the mid-1700s, being republished over and over again, the central story mutating multiple times as new copies were created, its message gaining rhetorical power as the historical context around it changed. This was the speech of Polly Baker.

To the modern eye, Polly Baker's narrative seems purpose-built to go viral—which, in an eighteenth-century kind of way, it did. First published on April 15, 1747, in the London *General Advertiser*, it claimed to be a transcript of a speech Ms. Baker had given at her trial across the ocean, "at Connecticut near Boston in New-England." Baker was being prosecuted for having a bastard child; not only that, but this was the fifth time she had been before the court on such a charge. But rather than being ashamed, Polly Baker was forthright. How was it fair, she argued, that she had been convicted multiple times for having illegitimate children, but the fathers of the children had got off without even a rap on the knuckles? "I have brought Five fine Children into the World, at the Risque of my Life; I have maintain'd them well by my own Industry," she said. "I have hazarded the Loss of the Publick Esteem, and have frequently endured Publick Disgrace and Punishment; and therefore ought, in my humble Opinion, instead of a Whipping, to have a Statue erected to my Memory."

Her speech was so forceful, the preamble to the text in the *General Advertiser* told us, that not only did the court refrain from punishing her, but one of the judges was so moved by her words that he got married to her the very next day. It seems ready-made to be translated into the language of mid-2010s viral inspiration sites: *This Woman Shut Down a Court with a Powerful Speech about Slut-Shaming—and What Happened Next Will Amaze You.*

This was, very clearly, Good Content. And so the republishing machine of the British press got to work. The day after Ms. Baker made her debut in the *General Advertiser*, at least five other London newspapers ran the speech, as well. It spread to papers in other cities: Northampton, Bath, Edinburgh, Dublin. A few weeks later, the news magazines, with their longer lead times, published it too. (None of them, fairly obviously, had had the time to pop over to Connecticut to see if they could track Polly down; the geographical effort barrier once again providing excellent cover for untruth to spread.) Not only did it get copied, but changes to the text started to appear, whether through error or intent—the most notable of these being in the *Gentleman's Magazine*, which decided that her simply marrying a judge wasn't enough of a plot twist, and they should have fifteen children together, as well. Exactly *when* those fifteen children were supposed to have been birthed was unclear, as the text was ambiguous about when the events had taken place.

A few months later, in July, the story had crossed the Atlantic and found its way into the nascent newspaper market of the American colonies, making its debut first in Boston, before migrating down the coast to New York and Maryland. Despite it being at least a little bit easier for the American press to look into the tale's authenticity, it's not clear that anybody bothered—which, honestly, is hardly surprising. Even in the age of telephones and Google, trying to establish that something *didn't* happen can be remarkably gnarly. At a time when there were only twelve newspapers in the country and the idea of the intrepid investigative journalist was still over a century away, it's perhaps unsurprising that they maybe felt they had better things to do. It was perfectly common practice to republish material from Britain's more developed press; the effort barrier to checking may have been reduced, but it got replaced with the assumed authority bestowed by the reputation of the British press. It was several of those structural problems rolled into one: there was a lack of imagination in the assumption that the British press must be reliable, which fed into a bullshit feedback loop on a grand scale.

So, rather than spurring on any further action (whether to delve deeper, or to debunk), Polly's rallying cry against sexist double standards quietly made its way into the canon of the collective consciousness, becoming one of those old favorites—the stories that form the background hum of the public psyche, whipped out every now and again when somebody wants to make a point. For the next few decades, it would crop up again and again; it was republished in newspapers, magazines, books; it was translated into Swedish and French. As a symbol of an ordinary person making a stand against unjust laws, it became big in the world of deism—the theological movement that argued against an interventionist God and the arbitrary authority of the Church—which would be a major intellectual influence on both the French and American revolutions.

It was in this context that Polly's speech got its second great lease of life, more than two decades after it was first published—which also led to the true story behind it being finally unmasked. In 1770, the anecdote appeared in a newly rewritten and far more melodramatic form, in a bestselling French history book by the Abbé Raynal—an ex-priest with a shaky grasp of history but a certain skill with agitprop. (At least, some of it was written by him; large sections were contributed by the somewhat more talented philosopher Denis Diderot, alongside a host of other collaborators. It's quite possible that Diderot was the one who added Polly's story to the text, as he seems to have been a fan of it.)

In the febrile atmosphere of prerevolutionary France, Polly Baker's oppression at the hands of tyrannical New England lawmakers struck a chord and became wildly popular. Raynal's history was reprinted multiple times in authorized and unauthorized editions, and other French versions of Polly's story appeared in the 1770s and 1780s. Which is how it came to be that, one day in 1777 or 1778, as the American Revolution was in full swing, Raynal paid a visit to America's minister in France, only to find him discussing the abbé's popular history book with a visitor from Connecticut.

None of the three people at this meeting ever wrote about what went on in the room where it happened. Instead, we have the story secondhand, via the then future American president Thomas Jefferson, who said he was told some years later what had played out that day. As with much of history, a pinch of salt is required.

The rough outline is this: the two Americans were discussing Raynal's book, and how bad it was, when, unexpectedly, Raynal walked in. The Connecticuter, one Silas Deane, greeted Raynal by cheerfully telling him they had just been chatting about how many errors were in his book. (Side note: as an author, I beg you, please do not do this—it's just rude; give it at least a couple of minutes of small talk first.) Raynal protested that there were no errors and that he'd been extremely careful to make sure that every fact in the book was authoritatively sourced.

"But what about Polly Baker?" asked Deane. "That's in there, and it definitely never happened."

"On the contrary," Raynal insisted, "I had an unimpeachable source for that, although I can't quite remember off the top of my head what it was, right now."

At this point, the American minister—one Mr. Benjamin Franklin—found himself unable to control his laughter anymore.

That was because it had been he who created the entire tale of Polly Baker, three decades earlier, and planted it in the British press. His career in faking hadn't ended with his untimely declaration of Titan Leeds's death.

In fact, it hadn't begun there either.

Franklin actually started his career of deceit in the news industry as a teenager, in 1722, when his older brother, James, banned him from writing for the *New-England Courant*, the newspaper James published. Pissed off at this stifling of his creative powers, young Benjamin did what any enterprising sixteen-year-old would do: he invented a middle-aged widow named Silence Dogood and submitted articles under her name. (Something you'll know if you've watched the important documentary film *National Treasure*, starring Nicholas Cage.) James Franklin, completely oblivious to their true author, published fourteen of these letters, and Ms. Dogood attracted quite a following, including several offers of marriage.

His first foray into perfidy being such an emphatic success, Franklin cheerfully carried on where Silence Dogood had left off. By 1730, he was publishing his own newspaper in Philadelphia, the *Pennsylvania Gazette*, in which he printed an entirely fictional account of a witch trial. In reality, there hadn't been any notable witch trials in America for several decades. He then moved on to *Poor Richard's Almanack* (once again writing in the voice of an invented character), where he killed the unfortunate Mr. Leeds.

To give you an idea of just how much effort Franklin put into even his pettiest hoaxes, in 1755 he printed and inserted an entire fake chapter into the Bible (the very much nonexistent fifty-first chapter of Genesis) simply so he could win an argument with a posh English lady.[48]

Polly Baker hadn't really been intended to foment revolution-

ary fervor; she'd been created mostly for Franklin's own amuse-
ment. It was joke that had just...got a bit out of hand.

This was all happening in the early years of the mass media;
the kind of news industry that we're used to wouldn't properly
emerge for several decades. And yet there were still many elements to it that we recognize today: the unthinking republishing of news without checking its veracity, the audience's uneasy
mixture of distrust and credulity, the way that a story that's too
good to be true will thrive regardless. And all of those things
would continue as the news industry continued to expand into
the kind of belching content behemoth we're familiar with in
our time. That's what we'll look at in the next chapter—where
we'll discover that Polly Baker was far from the only time that
a joke got a bit out of hand.

3

THE

MISINFORMATION

AGE

I

In New York City, in early August of 1835, fans of the news would have found plenty to talk about.

There was the weather, naturally—a sweltering heat that barely let up all month. There was a serious fire in Lower Manhattan. There was an increasingly tense political climate, dominated by the topic of slavery and the often violent clashes between Whigs and Democrats, in a year that had already seen the nation's first failed presidential assassination attempt. Among the more scientifically minded, there was giddy anticipation around the predicted imminent return to the skies of Halley's Comet. And there was a curious exhibit at the popular entertainment venue Niblo's Garden, put on by an ambitious young chap looking to launch a career as a showman—one Phineas

Taylor Barnum—which had caused a sensation since it opened on August 10.

Not only was there a lot of news, but the sheer *availability* of news was itself a major new development. The city had seen an explosion of penny newspapers being founded in the previous two years: a new class of affordable, mass-market publications, all aggressively competing for stories and for readers.

So, yes, there'd have been plenty to talk about in early August.

By the end of August, the only thing anybody was talking about was the race of bat people who lived on the moon.

It's important to point out (because I wouldn't want you to get the wrong end of the stick, here) that the red-haired bat people of the moon were not alone in the lunar landscape. Don't be silly. As everybody knows, they were part of a vibrant and complex space ecosystem that included—among other things—giant bipedal beavers tenderly carrying their children in their arms, high-speed spherical amphibians that rolled along the beaches of the moon's vast and bountiful rivers and lakes, and herds of small blue goat-faced unicorns that frolicked playfully among bucolic rolling meadows of scarlet flowers.

These celestial wonders were first revealed, gradually, over the course of a week at the end of that August to the readers of the New York *Sun*, which reprinted the news of their discovery for its American audience from an account that first appeared in a supplement to the *Edinburgh Journal of Science*. They were based on recent observations that had been made across the ocean, at the Cape of Good Hope, in South Africa, with a remarkable new telescope of unprecedented power and clarity, by the great astronomer Sir John Herschel.

The reports of the lunar findings sent shockwaves both through the city and around the world. They drew vast crowds to the paper's offices, sent rival newspapers scrambling to reprint the news, and dominated both popular conversation and popular culture, including spawning a wildly successful play at the Bowery Theatre that premiered less than a month later. And they helped confirm the *Sun*—a paper that had been founded only two years previously—as quite probably the biggest-selling newspaper in the world.

But (and please brace yourself, here, for a shock) *none of it was actually true.*

I know, I know. This is quite a lot of information to get your head around in a short space of time. But please believe me when I tell you that scientists have checked very carefully, and there are in fact no ginger bat-people living on the moon *at all*. Also, no goat-unicorns.

A French print by the Thierry brothers showing the bat people of the moon.

The Great Moon Hoax of 1835 was not, as the initial stories in the *Sun* claimed, the work of a "Dr. Andrew Grant," who had been "for several years past the inseparable coadjutor of the younger Herschel,"[49] but of a young English immigrant to the United States named Richard Adams Locke. Locke had been hired as the editor of the *Sun* just two months earlier. He wasted little time in making an impact.

If you had to point to a single period in history to identify the birth of the modern news industry, the middle of the 1830s in New York would be as good a candidate as any. Newspapers before this time were very different from the ones we buy today (or, more accurately, the ones we don't buy, but sometimes read the websites of). For starters, they weren't far off being luxury items, targeted exclusively at the wealthy merchant and political classes, with little attempt made to appeal to a broader audience. New York's existing newspapers, at the start of the 1830s, cost six cents apiece, well outside the range of affordability for most of the city's rapidly growing population. Made up of a single folded sheet of paper, they had only four pages, and both the first and last of those—the most valuable real estate for any modern newspaper editor—were given over entirely to a plethora of short advertisements, printed in dense columns of almost unreadably small type.

Thanks to the twentieth-century innovations of Rupert Murdoch, the words "page 3" are inextricably linked in the minds of British newspaper aficionados with photos of topless women. By contrast, page 3 of New York's newspapers in the early 1830s tended to have long lists of stuff like currency exchange rates and the details of ships newly arrived in port—the kind of information that was vital to merchants but virtually useless as soft-core erotica, unless you had some particularly niche kinks. Any actual news stories were relegated to page 2, a page that any modern newspaper journalist will recognize as "the place you put stuff that people don't read."

None of this particularly screams, "Please buy this newspaper!" But the somewhat unenticing format wasn't really a problem for the sales of these papers, because they tended to rely on subscriptions rather than newsstand sales (which was handy, given that there were no newsstands then). They also relied heavily on patronage—specifically, political patronage. This was the tail end of the "party press" era in the United States, when most news outlets were either owned outright by political partisans or relied on favors from their chosen politicians, like being granted lucrative government contracts in return for their unwavering and full-throated support.

This resulted in what could generously be described as "a vibrant and passionate popular debate on the great political issues facing a young country," or, slightly less generously, as "a bunch of egomaniacs talking shit about their rivals with no regard for accuracy."

This, uh, "passion and vibrancy" frequently spilled over into real life. The New York of 1835 was very different from the gleaming metropolis of today. There were no glass skyscrapers, naturally; instead of skyscrapers, they had feral pigs roaming the feces-covered streets. But, nonetheless, the city did have some characteristics that would be very familiar to today's New Yorkers: it smelled like hell in the summer; it didn't have a working subway; and it had a small but influential coterie of media professionals who took their interpersonal dramas way too goddamn seriously.

Newspaper editors were very closely identified with the outlets they oversaw, not least because they wrote the vast majority of the copy in their papers themselves. The distinction between the modern roles of the reporter (whose job is to go out and find the news) and the editor (whose job is to sit in an office demanding pictures of Spider-Man) were still kind of fuzzy at this point. As a result, the wild beefing between the partisan outlets was frequently deeply personal—and it was fairly common for rival newspaper editors, when they bumped into each other in the street, to simply beat the crap out of each other. One editor even took to carrying pistols with him after being physically attacked by the same competitor three times in one week.[50]

It was in this pungent atmosphere that the New York *Sun* rose in 1833, and changed the game forever. The idea behind the *Sun* (and the other pioneering outlets of the new "penny press" era) was a radical one: instead of charging the standard six cents, it would cost just one cent. Rather than relying on subscriptions and patronage, it would be independent, sold on the streets by a bevy of newsboys shouting the day's most dramatic headlines. As such, it would make the bulk of its money from advertising, which could suddenly reach a much wider audience thanks to the paper's dramatically higher sales. This wasn't news as a niche, high-end product sold to a small, homogenous in-crowd—this was mass-market, popular and populist, ready to talk to a wide range of readers...and reliant on eyeballs.

In other words, it had hit on the broad-strokes business model that large chunks of the news industry would follow for much of the next 170 years—pretty much up until the last few decades, when a combination of asset-stripping hedge funds and the internet came along to ruin everybody's lavish expense accounts. (To briefly digress: more than a few people have recently suggested that the news industry right now is desperately scampering back to the previous models, either making subscription-funded products targeted at smaller, elite audiences, or becoming dependent on the patronage of influence-hungry oligarchs. Either way, fun times in Newsville!)

The *Sun*, quickly hitting on a formula that would stand the test of time—namely, that stories about crime, disasters and human drama drew eyeballs—saw its readership grow to unprecedented heights. In early August of 1835, it boasted of having sold 26,000 copies, far more than even *The Times* of London— almost certainly the biggest newspaper in the world, prior to the *Sun*'s arrival.[51] This may have been largely thanks to the terrible fire on August 12, a conflagration that razed large portions of the printing district in Downtown Manhattan. This was a two-for-one boost to the *Sun*'s sales: not only was it a huge and dramatic news story that people were eager to read about, but rather conveniently it also destroyed the printing press of the *Sun*'s nearest rival, another penny-press upstart named the

Morning Herald, which at that point had been publishing for just three months.

As such, the *Sun* was perfectly placed that August to create a new media sensation. And yet the story of the moon people began small: a brief paragraph on page 2 of the edition that went out on Friday, August 21, titled "Celestial Discoveries." It noted that, at the Cape of Good Hope, Sir John Herschel had made "astronomical discoveries of the most wonderful description, by means of an immense telescope of an entirely new principle."

This paragraph, it turned out, was just the teaser trailer. The full story began to appear the following week, on Tuesday, August 25. But even then, rather than leading with the most sensational aspects first, the *Sun* took the time to build the narrative slowly. The first day's installment was, frankly, a bit dull, consisting mostly of descriptions of how the "immense telescope," with its seven-ton lens, supposedly worked.

But this approach actually played in the *Sun*'s favor. Rather than the skepticism that leading with *HOLY SHIT THERE ARE BAT PEOPLE ON THE MOON* might have provoked, the sober recounting—presented as a reprint from the august publication the *Edinburgh Journal of Science*—lent the tale an air of credibility and kept the readers coming back for more.

The following day, the *Sun* began to unveil the moon's wonders. Wednesday's installment revealed that the moon was populated with bountiful plant and animal life—including those fields of red poppy-like flowers, the rolling amphibious creatures and the blue goat-unicorns. This was remarkable enough but was nothing compared to day three, which announced the discovery of the upright beavers—animals who were clearly possessed of a degree of intelligence, who carried their young in their arms "like a human being," and lived in huts "constructed better and higher than those of many tribes of human savages."

By this point, the story was already a sensation, but the fourth installment, which dropped on Friday, August 28, took it to new heights. That was when the *Sun* introduced the world to the lunar bat-people: "Vespertilio-homo, or man-bat," as Herschel allegedly named them; creatures of about "four feet in height," with "short and glossy copper-colored hair" and yellow faces described as "a slight improvement upon that of the large orang outang." And, crucially, wings "composed of a thin membrane… lying snugly upon their backs, from the top of their shoulders to the calves of their legs."

Not only could these humanlike creatures fly, but they were clearly highly intelligent—they were "evidently engaged in conversation" and "their gesticulation…appeared impassioned and emphatic." And, just in case that wasn't enough to get people's attention, the article also noted, "Our further observation of the habits of these creatures, who were of both sexes, led to results so very remarkable, that I prefer they should first be laid before the public in Dr. Herschel's own work… They are doubtless innocent and happy creatures, notwithstanding that some of their amusements would but ill comport with our terrestrial notions of decorum." A section describing those "amusements" was rather ostentatiously censored out.

Yep. While the text tiptoes around saying it, the reader could be left in little doubt: my dudes, the bat people of the moon *fucked.*

The final two installments couldn't help but be a slight anticlimax after Friday's revelations, but nonetheless they managed to sustain the by now almost-insatiable reader interest. Saturday brought the discovery of great, mysterious, temple-like buildings built from sapphire on the moon, while (after taking a break on Sunday) the following Monday introduced a new and improved variety of bat people. Described as "the highest order of animals in this rich valley," shown sitting in circles, having conversations, these better bat-persons were said to be "of a larger stature than the former specimens, less dark in color, and in every respect an improved variety of the race."

That's right—the bat people of the moon were in existence for a grand total of four days before someone was extremely racist about them.

The *Sun*'s office was besieged by crowds of thousands of people demanding updates, and their printing press couldn't churn out new copies fast enough.

Not only were the crowds keen to learn more, they actively contributed to the hoax. William Griggs, a friend of the hoax's author, Locke, told of how he heard people in the crowd offer supporting evidence to back up the fiction, in a state of "insatiable credence." One "highly respectable-looking elderly gentleman, in a fine broadcloth Quaker suit" claimed to have seen Herschel's fictitious telescope with his own eyes as it was loaded onto a boat at London's East India Docks; another man "of perfectly respectable appearance" insisted that he owned an original copy of the report in the *Edinburgh Journal of Science*, and that the *Sun*'s reprint was faithful. Griggs describes these as acts of "spontaneous mendacity."[52]

Benjamin Day, the *Sun*'s canny publisher, knew when he was onto a good thing and immediately saw an opportunity to cash in. Before the series had even completed, he'd republished the text as a stand-alone pamphlet, which rapidly sold tens of thousands of copies (at twelve and a half cents each). He commissioned artworks depicting the inhabitants of the moon. And he would go on to invest in new steam-powered printing presses to ensure that the *Sun* never had to run short of copies again. News was on its way to becoming an industry.

That the hoax was widely believed seems beyond doubt. New Yorkers of the time wrote about it in their diaries, and few seem to have expressed skepticism; multiple accounts from contemporary sources state that most people were taken in by Locke's writing. No less a figure than Edgar Allen Poe would later write that "not one person in ten discredited it… A grave professor of mathematics in a Virginian college told me seriously that he had no doubt of the truth of the whole affair!"[53] Poe himself was, at the time, supremely pissed off about the hoax—not because he got fooled, but because he'd published his own hoax story about a trip to the moon a few months earlier in the *Southern Literary Messenger* and had been planning a sequel before the *Sun*'s work blew his out of the water.

Eventually, however, people started to express public skepticism—and among the first out of the gate was one James Gordon Bennett, the editor of the *Morning Herald*. He'd been forced to sit out the first week of the hoax, as his paper still wasn't able to publish following the fire earlier in the month, and presumably he had been grinding his teeth about the success of his rivals all that time.

Come the following Monday, August 31, however, the *Herald* was back in business (and had dropped the *Morning* from its name[54]). Bennett immediately launched into a broadside against his competitors with an article titled "The Astronomical Hoax Explained"—which noted, among other things, that the real *Edinburgh Journal of Science* had ceased publication two years earlier, and as such couldn't possibly be the source for the tale. He would continue in the same vein for weeks to come, calling the *Sun*'s actions "highly improper, wicked, and in fact a species of impudent swindling," and accusing them of printing "untruths for money."[55]

(In case you're wondering, James Gordon Bennett is not exactly where we get the phrase "Gordon Bennett" from, but he's not unconnected. It actually comes from his son, James Gordon Bennett Jr., who inherited his father's publishing duties at the paper that eventually settled on the name of the *New York Herald*, the title under which it found the greatest fame. Bennett Jr. was not simply a newspaperman, though: he was also so deeply committed to a wild, louche and very publicly eccentric lifestyle that his name became pretty much a synonym for "holy shit.")

Was his accusation accurate? Was the *Sun* telling untruths

purely for money? Okay, so, undoubtedly the pressure for sales helped to push Locke toward writing his *magnum hoaxus*, and Day, as publisher, certainly didn't hesitate for a second to wring every dollar he could out of the sensation. But Locke seems to have had other motives, as well. In fact, according to his own explanation, when (some years later) he finally confessed to the hoax, he had created one of history's more famous lies because he was himself annoyed about people spreading untruths. The piece was intended, not as a hoax, but as a parody of "natural theology," a popular philosophy of the time, in which science was relegated to second-class status in the quest to understand God's design. As a fan of science and an enthusiast for both geology and astronomy, this way of thinking appalled him. He wanted to show it up for the charlatanry it was.

Locke hadn't really meant to spread bullshit. He'd just told a very elaborate joke that almost nobody got.

It was a joke that backfired on Locke in unfortunate ways. For the rest of his life, he couldn't escape the shadow of his moon. He left the *Sun* a year later and moved to a new newspaper, where he hoped to do work of greater value to the world, but it failed. A few years into that gig, he tried another hoax, about the supposed lost diaries of the Scottish explorer Mungo Park, but nobody cared. His writing was suspect. Locke drank more and more heavily. Less than a decade after his hoax was published, he left journalism entirely and spent the remaining three decades of his life quietly working for the Customs Service.

But what he left behind has lasted to this day. The legacies of the Great Moon Hoax—newspapers battling for circulation, the industrialization of news distribution and the prioritization of sensation over accuracy—are factors that have resonated through the business of recording news for almost two centuries. In the words of that invaluable online resource, the Museum of Hoaxes, Locke's moon series was "the first truly sensational demonstration of the power of the mass media."[56]

And, like Franklin before him, it was a journalist's joke that just got out of hand. It wouldn't be the last time either.

I should probably, at this point, confess my own interest here: I am both a journalist and someone who has made jokes that have got out of hand. For many years, recently—thanks to the quirks of new media in general, and to the indulgence of my bosses at BuzzFeed in particular—my job combined both of those vocations into a single, somewhat confusing package. On the one hand, as a journalist, I was reporting about online viral misinformation: helping to expose an unscrupulous news agency, hunting down Russian bots, debunking what felt like an infinite number of photoshopped pictures of sharks swimming down flooded streets. On the other hand, as a humor writer, I was creating elaborate spoofs of media reports on nonexistent events.

Those spoofs, of course, were almost without exception interpreted as real by at least a small subsection of the readership. I learned to my cost that there is basically nothing you can do—short of perhaps writing *THIS IS A JOKE, YOU GUYS* all over something, which tends to slightly spoil the punch line—to stop someone, somewhere mistaking even the most thuddingly obvious joke for reality. If you're a fan of questioning your place in the world, there's not much quite like seeing a gag you created about people credulously sharing nonsense on Twitter being shared as a real thing on Twitter less than a year later.

As such, I would also like to take this opportunity to—for the first time—publicly apologize to the senior BBC presenter Nick Robinson, and to clarify that he did not go to Eton with David Cameron and was never secretly recorded saying, "I hate all poor people."[57] I cannot stress enough that *it was a joke and people weren't supposed to tweet it out of context.*

In addition to my lingering shame about traducing Mr. Robinson, this background also means that I have a somewhat divided view of journalism. I will, like all journalists, staunchly and somewhat pompously defend it as a noble and courageous profession, a vital pillar of any democratic society, and an essential tool for uncovering truth and holding the powerful to account. This isn't simply a posture; every day I am inspired by journalistic colleagues across the world, many of whom risk imprisonment, ruin or death to expose wrongdoing and shine a light in the darkness. They're heroes.

I am, however, also aware that quite a lot of what the news industry produces is—to varying degrees—rubbish.

Now, this is partly just because the job of finding out facts and writing about them to a tight deadline is actually quite hard. Not necessarily hard in a "working down a mine" kind of way; more, hard in a "trying to find a needle in a haystack, and also the haystack is in a tornado, plus nobody is actually 100 percent sure that the needle was even in the haystack in the first place, the farmer has started referring all questions about the haystack to his lawyer—oh, and the guy from Reuters got here two hours earlier and has already scored an exclusive with the needle's family" sort of way.

Put bluntly, human events are messy and chaotic, and trying to establish what really happened in even the most minor incident—and then distill it into eight hundred clear and crisp words, all within the space of a few hours—can honestly be fairly tricky sometimes.

Nowhere is this more apparent than in the 1904 story of a snake that appeared unexpectedly one day in a New York apartment.

To be very clear up front, this is not, in the grand scheme of things, an important story. No governments fell, no great social movements were sparked, its legacy failed to echo down the years. Precisely zero musicals have been written about it. The only victim in the tale was dead well before the story hit the printing presses—that victim being the snake in question.

The unfortunate snake made its appearance in an inauspicious apartment at Twenty-Two East Thirty-Third Street, in an insalubrious and crime-ridden district of Manhattan (decades later, after being turned into offices, the same address would briefly host the final iteration of Andy Warhol's Factory). There, a small boy was spotted by his family playing with an unusual-looking new toy. The new toy, a more detailed inspection revealed, was actually a live snake.

The family, rather naturally, freaked the fuck out, and very quickly hacked the snake to death—RIP snake—before carrying its corpse several blocks to the run-down, foul-smelling local police station. From where, we have to assume, one of New York's finest tipped off the gentlemen of the press that a quirky human-interest story was up for grabs.

You can see why this would be of interest to the newspapermen; it's the kind of story that local news coverage is made of. The reason this is of interest to *us*, however, is what happened next: none of the six newspapers that covered it could agree on a single detail of what had actually occurred.

We have to thank Andie Tucher, a former journalist and a historian at New York City's Columbia School of Journalism, for her exhaustive investigation of the dizzying array of inaccuracies the press managed to wring out of this completely inconsequential story.[58] Between the New York *Sun*, the *Herald*, *The Times*, the *Tribune*, the *World*, and the *American and Evening Journal*, conflicting details were sprayed around like ticker tape. They disagreed about the size of the snake (anywhere between three and five feet long), the color of the snake (yellow or brown or green or black, sometimes with spots in various color combinations), the age of the boy (three, four, five or none of the above), as well as the name of the boy (Pierre, or possibly Albert, Jeltrup or Gultrep, or Blanpain) and the name of the neighbor from whose menagerie the snake had supposedly escaped (while they agreed his first name was Gustave, his surname was given as Hurtiland or Svenson, neither of which was right). Furthermore, they disagreed about who killed the snake (father, grandfather, uncle, nurse), what implement they killed the snake with (knife, shovel, hammer, sword), and even how many pieces the snake was in after it had been killed (two, or many, many pieces).

Essentially, every possible detail of the incident beyond "there was a snake" was in dispute. It's like a weird herpetologist version of Clue.

The point of this is not to rag on the long-dead beat reporters of New York's nineteenth precinct but simply to highlight just how tenuous the connection between what was real and what was reported has been for much of our history. This was a story, after all, that featured very few of the problems inherent in more serious journalism: none of the subjects (with the possible exception of alleged snake-source Gustave) had any incentive to bend the truth. Nobody was trying to cover anything up, nobody was promoting a movie and nobody was trying to use the snake as political justification for the military invasion of a third world country.

Some of the reporters of the inaccurate snake story may simply have been lazy, or incompetent or simply unlucky. But, then again, they may have just been practicing their trade as best they knew.

Nowadays, the phrase "fake news" is pretty much everywhere—and has seen its meaning rapidly and depressingly shift from "completely fictional copy masquerading as news purely to drive clicks" (2016 meaning) to "stuff printed about a politician that the politician doesn't like" (2017 to present). But this isn't the first time the term "fake" has entered the news industry only to see its meaning morph over time; something very similar happened in the late nineteenth and early twentieth centuries, when the word first made its appearance in the world of news.

Before this time, the concept of "faking" wasn't part of the mainstream discourse; it was used as a term of art by only the most deeply disreputable professions, such as thieves and con men and actors. But—as the snake-investigating journalism historian Tucher has written[59]—by the late 1880s it had made its way over to the newly professionalizing world of journalism. Except it wasn't necessarily treated as the original sin of reportage—the kind of thing that would get someone drummed out of the industry. According to some authorities, it was seen as an essential job skill.

In *The Writer*, a magazine launched in 1887 for the burgeoning class of professional scribblers, the editor, William Hills, wrote approvingly of newspapers where journalists must be "able to 'fake' brilliantly to do the work well."[60] A few months later, he insisted that "hardly a news despatch is written which is not 'faked' to a greater or lesser degree";[61] he describes the act as "the supplying, by the exercise of common sense and a healthy imagination, of unimportant details…[that] may not be borne out by the facts, although they are in accordance with what the correspondent believes is most likely to be true." The purpose was simply to make the story more "picturesque"; faking, he insisted, was "not exactly lying."

An example of precisely what those "unimportant details" might be comes in an 1894 training handbook for young journalists, written by Edwin Shuman, a Chicago journalist who also taught a course on how to be a journalist, at a time before journalism degrees were a thing. Shuman warns these wannabe journalists against the "dull and prosy error of being tiresomely exact about little things, like the minutes and seconds or the state of the atmosphere or the precise words of the speaker."[62]

If you're a news editor, there's a decent chance that you might have just screamed out loud at those last six words. Granted, Shuman was writing at a time when you couldn't exactly slip a recording device into your pocket (back then, the machines that a decade later would be trademarked as the Dictaphone involved a rather bulky wax cylinder). But, still—the precise words of the speaker are not a "little thing"!

So faking was commonplace. It gave reporters (playing up to the stereotype of frustrated novelists) the chance to flex their literary muscles, and it was a useful way of avoiding being scooped, which was a far greater offense to the gods of journalism. Editors liked it because it ensured a steady flow of sparkling copy; readers liked the product and rewarded it with sales. If news stories that seemed a bit too good to be true continued to pour in—particularly from small or remote locations where the effort barrier to checking them seemed just too great—then nobody was going to do much to intervene.

Such was the career of Louis T. Stone, an ambitious young writer from the small town of Winsted, Connecticut, who quickly rose to become one of the most-read journalists in the country, thanks to the almost insatiable appetite of many newspapers for the dispatches he filed from his hometown. The "Winsted Liar," as he became known, enjoyed a decades-long career that spanned from 1895 to his death in 1933, during which he produced a steady stream of nonsense that editors couldn't resist.

Among Stone's more notable reports, as recorded by journalism professor Curtis D. MacDougall in 1940, were the following: a red, white and blue egg laid by a hen on July 4; a tree that grew baked apples; a cat that whistled "Yankee Doodle"; a watch swallowed by a cow that kept almost perfect time for years in the cow's stomach because the cow's breathing kept winding it up; and a bald man who painted a spider on his head to repel flies.[63]

That's quite an eyebrow-raising collection of tales at the best of times, but, when they all originated from the same writer in the same small town, you'd think someone would have caught on—either to Stone's fakery, or to the remote possibility that Winsted was a portal into fairyland. Did anybody actually believe them? MacDougall insists that they were taken as true by "virtually everyone...except wise editors who came to be skeptical of anything submitted by Stone but printed the stories anyway because of their reader interest."[64]

None of the skepticism affected Stone's career; he rose to become the general manager of his local paper, having turned down numerous offers of big-city work, preferring to stay in his small town, where the news could remain weird. When he died, rather than being disgruntled at his fictions, the grateful citizens of his hometown praised him for "putting Winsted on the map," and named a bridge in his honor—a bridge that crossed a local river named Sucker Brook.

Beyond a bridge and an eternal place in the pantheon of journalistic fakers, not many of Stone's fancies have left much of a legacy. But the same can't be said for one of the nineteenth century's most notable hoaxes: the letters that gave us the legend of Jack the Ripper.

The Whitechapel murders gave us one of the most enduring pop-culture figures of our time, and if it feels particularly grim to describe a suspected mass killer as a "pop-culture figure," then...yeah. The deaths have been the inspiration for films, TV series, novels, songs, comics, exhibits and at least one fictional musical. On some weekends, it's almost impossible to walk around certain parts of East London without finding your path blocked by "Ripper tours," in which precariously employed actors intone spooky stories about dimly lit streets and shadowy figures in the smog to a crowd of eager tourists, all doing their best to ignore the fact that the pub they're standing outside is now full of advertising creatives doing vape tricks.

But a large amount of what most of us "know" about Jack the Ripper and his victims (quite possibly up to and including the belief that there was definitely a single serial killer responsible for the five "canonical" murders) is based on a slightly hazy mixture of truth, supposition and contemporaneous reporting that wasn't going to let facts get in the way of a good tale. That includes much of the core mythology of Jack the Ripper himself—most notably, his nickname.

The name "Jack the Ripper" stems from three communications that were supposedly received by the Central News Agency in September and October 1888: the "Dear Boss" letter, the "Saucy Jacky" postcard and the "Moab & Midian" letter. Written in red, they were signed "Jack the Ripper" ("Dont mind me giving the trade name") and helped set the template for every airport-thriller serial killer taunting the police that was to follow. They gave a motive ("I am down on whores") and the promise of future atrocities ("My knife's so nice and sharp I want to get to work right away if I get a chance"), they mocked the police for not apprehending the culprit ("I keep on hearing the police have caught me but they wont fix me just yet"), and they described the grisly keeping of souvenirs ("I saved some of the proper red stuff in a ginger beer bottle").

The modern consensus is that these communications were almost certainly the work of a journalist looking to keep the story going. Both handwriting and linguistic analysis suggests that they were written by the same author, and the finger is generally pointed at either Fred Best (a freelancer who supposedly confessed to being the author decades later, although the source of that is more than a little dubious) or Thomas Pulling, who actually worked at the Central News Agency and was responsible for forwarding the letters to the police—and, somewhat suspiciously, only forwarded a transcript of the final letter, with the original nowhere to be found, rather firmly placing it into the "likely hoax" category.

From the turn of the twentieth century onward, the practice of casual faking gradually came to be frowned upon by the increasingly professionalized journalism industry—but that doesn't mean it went away. The modern history of journalism is littered with prestigious reporters who made up most or all of their work, each time unleashing a torrent of soul-searching within the industry and promises that it could never happen again. Many of the names are familiar: Jayson Blair; Stephen Glass; Janet Cooke, who won a Pulitzer in 1981 for a fabricated story of an eight-year-old heroin addict called Jimmy. In December 2018, the German magazine *Der Spiegel* fired its award-winning journalist Claas Relotius, who had been filing fictional copy from the troubled faraway lands of the USA, once again relying on a bit of geographical distance to make checking the lies more difficult.

Again, some of these false stories have irrevocably entered our cultural consciousness. *Saturday Night Fever* remains an iconic moment in both film and music, something that hasn't changed much since Nik Cohn—the Northern Irish music journalist who wrote "Tribal Rites of the New Saturday Nights," the *New York* magazine article it's based on—admitted he made the whole thing up. As Cohn tells it, he got a taxi out to Bay Ridge in Brooklyn, planning to write about the vibrant disco culture he'd heard was thriving at a club there. The moment he opened the cab door, a man who had been having a fight in the street vomited down Cohn's leg—whereupon the journalist immediately shut the door and fucked off back to Manhattan, deciding to invent the details of this vibrant disco culture instead. John Travolta's character, in all its gritty depiction of working-class Italian-American life, is actually based on a mod called Chris whom Cohn had met in London a decade earlier.

Cohn told the *Guardian* in 2016 that he was surprised, but not shocked, that it got published: "It reads to me as obvious fiction… No way could it sneak past customs now. In the 60s and 70s, the line between fact and fiction was blurry. Many magazine writers used fictional techniques to tell supposedly factual stories. No end of liberties were taken. Few editors asked tough questions. For the most part it was a case of 'don't ask, don't tell.'"[65]

It's not always a question of journalists making stuff up, though. Sometimes, it's the newspapers who are getting hoaxed—as with the press's apparently insatiable appetite for stories about Nazis that aren't true. The most famous of these is of course the Hitler Diaries from 1983, forgeries that were the work of a petty criminal and smuggler of Nazi memorabilia, which nonetheless fooled such august publications as the German magazine *Stern* and *The Times* of London, as well as the acclaimed historian Hugh Trevor-Roper. But it would be unfair to overlook the *Daily Express*'s publication in 1972 of the sensational news that it had "incontrovertible evidence" that Hitler's deputy Martin Bormann had been found living somewhere in Latin America. The incontrovertible evidence turned out to be a photo of a man who was not, in fact, a Nazi in hiding, but an Argentinian schoolteacher. The *Express*'s source for the story— an unreliable Hungarian-American historian and Nazi hunter with the unimprovable name of Ladislas Faragó—didn't let a little thing like that stop him from publishing a book about his search for Bormann, two years later.

The history of newspapers getting hoaxed is a long one. A particularly entertaining example comes (again) from *The Times*, in October 1856, when London's newspaper of record published a shocking tale of violence from the state of Georgia in the USA. It describes a lengthy train ride during which, following a series of altercations, at least five deadly duels were fought, leaving six dead once the smoke had cleared. "They fought with Monte Christo [sic] pistols, or pistols that make no report," *The Times* solemnly outlined, lamenting the nightmare of savagery that the former colony's southern states had descended into. "Of the six killed two were fathers and two were their sons, one father killed while avenging his son, and one child murdered for lamenting his father," it wrote—a little boy apparently having had his throat cut because he wouldn't stop crying.[66]

The atrocity, *The Times* wrote a day later (while insisting that the story was undoubtedly true), should prompt "some rather serious reflections as to the future of the United States, for what we have described appears to be the 'normal' state."[67]

When this report reached America, a little over a week later, it would be fair to say that the US press were not having it; the tale of the Monte Cristo pistols sparked a right old transatlantic slanging match. The Augusta *Constitutionalist* and *Chronicle & Sentinel*, and the Savannah *Republican*, were among the papers to lay into *The Times*'s "superlative gullibility" in publishing the "utterly preposterous" tale.[68] "A Prodigious Hoax," the *New York Times* labeled the work of its London namesake, with its editor writing that the story was "so full of grotesque nonsense and incredible absurdities that we think that the editor of the Times instead of vouching for the sanity and truthfulness of the narrator, should have furnished some evidence of his own sanity in admitting such a farago of nonsense into his columns."

The dispute rumbled on for months, back and forth over the sea, becoming a major political issue. For much of that time, *The Times* stood steadfastly behind its story, insisting that it was true, before finally—after a British consul forwarded them a disgruntled letter from the president of the Central Georgia Railroad—coming to acknowledge, in mid-December, that maybe their source might have been "hallucinating."

The punch line of the whole affair only arrived in the summer of 1857, however, when *The Times*'s own special correspondent in the United States, a chap named Louis Filmore,[69] managed to track down some more details of the story. As he happened to be passing through the area, he interviewed some train passengers, who all denied that any such incident had ever occurred. The tale was a "monstrous tissue of fiction," Filmore wrote. But his interviewees in the baggage car of the train (which doubled as the favored location for smoking and taking refreshments) did reveal one important detail: that "Monte Christo pistols" was, in fact, local slang for bottles of champagne, and empty bottles of the same were known as "dead men." Filmore noted, deadpan, that "encounters with the Monte Christo weapon in the baggage wagon are, I understand, not uncommon on the line."[70]

Of course, it doesn't require a deliberate intent to deceive for things to get a bit out of hand. Most of the examples we've seen so far have involved outright hoaxes, or at least some will-ful embellishment of a story. But sometimes there's no hoax at all—and yet a basically true article can nonetheless get blown out of all proportion, as newspaper after newspaper adds a little bit of extra sensationalism to the story each time it's published.

That's what happened in 1910 when, once again, the news-papers of New York turned their gaze heavenward as Halley's Comet made its first appearance in our skies since those heady days of 1835—and a perfectly accurate report in the *New York Times* triggered an apocalyptic panic.

The report was a mere three paragraphs, halfway down page 1, with the simple headline of "Comet's Poisonous Tail."[71] It re-layed that astronomers had, using new spectroscopy techniques, discovered that the tail of Halley's Comet contained a signifi-cant quantity of cyanogen. Cyanogen, it reminded us, is "a very deadly poison," and the article reported that the discovery had prompted "much discussion" among astronomers over what ef-fect this would have on the earth "should it pass through the comet's tail." At the end of the second paragraph, the *New York Times* just casually drops in the opinion of a French astrono-mer named Camille Flammarion, who believed that it would "impregnate the atmosphere and possibly snuff out all life on the planet."

In journalism circles, that's known as "burying the lede."

The thing is, this was a perfectly accurate report of Flam-marion's opinion. The *New York Times* even added, in the very next paragraph, that "most astronomers do not agree with Flam-marion," but that wasn't enough. The idea was out there now, and if there's one thing humans know how to do well, it's panic for no good reason.

As Halley grew closer, so did the fear of impending doom. Reports from the time tell us that people were blocking up their windows and doors to keep out the poisonous fumes; sales of gas masks were brisk; some con artists even took to selling anti-comet pills that, they claimed, would protect the public from the effect of the comet's deadly tail. Another report from the *New York Times* on May 19, about the reaction in Chicago, is headlined "Some Driven to Suicide," with the subheading "Others Become Temporarily Insane from Brooding Over Comet."[72]

In the end, the comet passed by without any deadly effects, with the exception of a sixteen-year-old girl who fell off a roof in Brooklyn while having a comet-watching party.

It's this ability to blow things entirely out of proportion, and the dogged refusal to let go of a notion once it's fixed, that are at the heart of how the press gets things wrong. Even without any deliberate falsification, the collective hive mind of the press—bolstered by reader feedback on what they like to read about—will often center itself on a particular idea from which they cannot easily be shaken. This framing of what's going on, the narrative that becomes "the story," has incredible momentum once it gets going—and anything that is not "the story" struggles to see the light of day.

British readers may remember, for example, the terrifying recent tale of the Croydon Cat Killer, a maniac who was supposedly responsible for killing and mutilating hundreds of cats in the Croydon area of South London. Reports of the Croydon Cat Killer first emerged in 2015, after some concerned residents in the area went to the *Daily Mail*. The rest of the press jumped on the story. When would the Croydon Cat Killer strike next? Why weren't the police doing more to catch him? We were warned that it was only a matter of time before the sadistic killer turned his attention from felines to humans.

Examples of dead cats from farther afield than Croydon were taken, not as evidence that there are a lot of cats in Britain and sometimes they die, but that "the cat killer—who is believed to have killed more than 100 moggies in the past year—is now thought to be operating on 'much-wider scale.'"[73] He was renamed the M25 Cat Killer; by the time his killing spree had traveled as far as Manchester, he basically just became the Cat Killer. The story was a central obsession of much of the British press for well over a year.

Eventually, in September 2018, the Metropolitan Police announced that they had found the culprit—or, rather, the culprits. The Croydon Cat Killer was...cars and foxes. That's it. It was just cats who'd been run over by cars, and had sometimes been chewed by foxes after their deaths. It took over two thousand hours of police time and at least $7,000 worth of cat autopsies to establish that, however.[74]

This ability to create something out of nothing, and for the narrative to then snowball until it becomes unstoppable, is hardly new. The saga of the Croydon Cat Killer is remarkably similar to that of the Mad Gasser of Mattoon from seven decades earlier, in which one small piece of melodramatic wording in a single news article led to a weeks-long panic across the quiet city of Mattoon, Illinois.

The story, at its core, was fairly simple. On September 1, 1944—while the Second World War was at its peak and fears of Nazi attacks were commonplace—a woman named Aline Kearney thought she smelled an unusual odor and shortly afterward suffered some kind of episode, feeling faint and complaining of a paralysis in her legs. The police were called and found nothing suspicious, while she recovered within half an hour. But when her husband returned home about an hour later, he thought he saw a figure lurking near the house, although again the police were unable to find any intruder.

The following day, the *Mattoon Daily Journal-Gazette* ran it as their top story, with a bold front-page splash headline. "'Anesthetic Prowler' on Loose," it shouted, adding, "Mrs. Kearney and Daughter First Victims."

You can see what they did there, right? Not only did they take a vague suspicion and turn it into a specific plan—an intruder silently filling a house with anesthetic gas, so that he could break in when the occupants were unconscious—but with the addition of a single phrase, "First Victims," they primed every single person who read it to expect further attacks.

It was, naturally, a self-fulfilling prophecy. Anybody who'd come over a bit faint in the previous weeks suddenly suspected that perhaps they, too, had been an early victim of the Gasser; their stories, eagerly reported by the *Daily Journal-Gazette*, served only to bolster the certainty that there was a dangerous individual on the loose. Further reports came in over the following days. Within a week, other newspapers in the region had picked up the story, all taking it as a point of certainty that the original reports had been valid.

The headlines kept coming: "'Mad Anesthetist' Strikes Again! Visits 2 More Homes in City during Night"; "'Mad Gasser' Adds Six Victims! 5 Women and Boy Latest Overcome." The reports were every bit as sensational as the headlines. On September 10, the *Chicago Herald-American* described the scenes in Mattoon: "Groggy as Londoners under protracted aerial blitzing, this town's bewildered citizens reeled today under repeated attacks of a mad anesthetist who has sprayed deadly nerve gas into 13 homes and has knocked out 27 known victims."[75]

Even on days when no new reports came in, it was still evidence that the underlying story was real: "Mad Prowler Takes It Easy for Night," ran that day's *Daily Journal-Gazette* headline.

By this point, panic was virtually universal across the town. Crowds surged into the street when someone said they'd spotted the Gasser; naturally, someone then smelled an unusual smell and many in the crowd became convinced they'd been gassed. Some people were hospitalized. The small local police force was overwhelmed.

It was only at this point, a week and a half after the first attack, that the authorities—who previously had accepted the reports as real—began to push back, openly describing the fears as "mass hysteria." And now that "the story" had changed, so did the behavior of the press; they began to openly mock the panic and interviewed psychologists to explain how people had fallen for the "gasser myth." In an impressive piece of blame judo, responsibility for the mass hysteria was pinned on chemical gasses from nearby factories.

The role of the press in stoking the panic was conveniently overlooked.

Ultimately, all this would be of limited consequence if newspapers really were, as the old cliché goes, tomorrow's fish-and-chip wrapper. But they're not. What the press says has a tendency to stick around. As the other old cliché goes, journalism is the first draft of history. The only trouble with this is that, regrettably often, nobody bothers doing the second draft for *ages*, if at all.

And we can see that worryingly well in the tale of another journalistic joke that—you're there ahead of me—got a bit out of hand.

The Great Moon Hoax may be the most famous falsehood in the storied history of American journalism (for unclear reasons, under American journalism rules, it must always be a "storied history," never just a regular fuckin' history), but it's challenged for that crown by the work of someone who, in the *extremely* storied history of US newspapermen, stands as one of the more notable figures: H. L. Mencken.

Mencken was one of the most lauded writers and editors in the first half of the twentieth century, a wry and frequently savage commentator on politics and society at large, a man described by the *New Yorker* as "the most influential journalist that America ever produced."[76] A quote from him appeared for several decades in big letters on the office wall of his former employer, the *Baltimore Sun*. (You can see the wall quote in the confusingly unsubtle final season of *The Wire*, which is about journalists making stuff up; in 2018, shortly after the paper moved offices, the *Sun* noted on Twitter that the quote had given the wrong date the entire time it was up.[77])

Let us also note for the record that Mencken was a complete asshole: a pissy contrarian, an elitist snob and, above all, just a massive, massive racist. Hated poor people, hated black people, hated Jews. This doesn't particularly affect what comes after this, but it's worth mentioning, especially as the other two journalistic hoaxers who get starring roles in this chapter were honestly pretty decent people, by the standards of their time. Mencken was not. Horrible man; lovely turn of phrase.

Anyway. In December 1917, as the First World War raged, Mencken published a gentle and amusing column about the history of bathtubs in the United States, to mark what he described as the "neglected anniversary" of the bathtub first coming to America—the pioneering tub having been installed by an enterprising merchant named Adam Thompson, in his Cincinnati townhouse, in December 1842.

The column was (as Mencken would despondently confess, eight years later) "pure buncombe" and "a tissue of absurdities."[78] There was no Adam Thompson; he had not been inspired by Lord John Russell's introduction of the bathtub to England in 1828 (which wasn't true either); and Americans had not belatedly come around to the idea of baths only after President Millard Fillmore controversially had one installed in the White House (this was also bull).

Mencken wrote the piece simply as a joke, "to relieve the strain of war days." As a fervent Germanophile and an opponent of the USA's entry into the First World War, he found himself in possession of an unpopular opinion that he was constrained from writing, and he was increasingly grumpy about what he saw as newspaper reports from the war that were full of falsehoods. As he wrote later of the war years, "How much that was then devoured by the newspaper readers of the world was actually true? Probably not 1 per cent."[79] (Which is perhaps harsh on the journalists covering that war—but not entirely off base, as we'll see in a later chapter.)

The bathtub hoax was merely Mencken's way of letting off a little steam. Unfortunately, he did far too good a job. The article was filled with myriad details that lent it a superficial but delightful plausibility; it had the bouncy, slightly drunken zigzag gait of authentic history. The reader was told that Thompson's bathtub was supposedly made from "Nicaragua mahogany," lined with lead and "weighed about 1,750 pounds."[80] The bathtub sparked immediate controversy, with fears that it would cause "phthisic, rheumatic fevers, inflammation of the lungs, and the whole category of zymotic diseases"; bathing was almost banned in Philadelphia and Boston; Virginia introduced a bathtub tax. Political opponents attacked President Fillmore for his decision to introduce a bathtub to the White House, claiming that his bathing-centric actions seemed disconcertingly French.

Mencken was initially pleased with the column, but (as he wrote in his 1926 confession, titled "Melancholy Reflections") very quickly his "satisfaction turned to consternation." *People hadn't realized it was a joke.* Other newspapers reprinted the article, or rewrote it. Readers started writing to him, taking his article seriously, even offering supporting evidence for his entirely fictitious history—another example of William Griggs's "spontaneous mendacity."

"Pretty soon I began to encounter my preposterous 'facts' in the writings of other men," Mencken continues. "They got into learned journals. They were alluded to on the floor of Congress. They crossed the ocean, and were discussed solemnly in England and on the continent. Finally, I began to find them in standard works of reference."

Mencken's admission that the story was entirely false was published in around thirty newspapers on May 23, 1926. The column remains a classic in the field of bullshit studies, with its pointed observations of the news industry's fallibility. Mencken writes: "As a practicing journalist for many years, I have often had close contact with history in the making. I can recall no time or place when what actually occurred was afterward generally known and believed. Sometimes a part of the truth got out, but never all. And what actually got out was seldom clearly understood."

Which, all told, was a fairly apt summary of the state of things. But what Mencken couldn't have known at the time he wrote those words was just how powerful an example of this business the whole saga would become. Because the most remarkable thing about Mencken's bathtub hoax isn't that the original spoof was believed, or that people then began to repeat it. As we've seen, that pretty much comes as standard.

No, what makes this a banner moment in the history of bullshit is this: an admission from the author in the pages of multiple newspapers that it was all lies *did absolutely nothing to stop its spread*.

Incredibly, despite Mencken's belated attempt to put this particular genie back in its bottle, the tale of Adam Thompson's pioneering bathtub simply refused to die. People just kept on repeating the factoids from it, oblivious or uncaring about the fact it was bunk.

Writing a decade after Mencken's front-page confession, the not-especially-good Arctic explorer Vilhjalmur Stefansson recorded in his book *Adventures in Error* an incomplete list of more than thirty times the false story of the bathtub had been repeated by prominent sources in the ten years since Mencken's admission.[81] These included American papers such as the *New York Times*, the *Baltimore Evening Sun*, the *Cleveland Press* and the *New York Herald* (multiple times); foreign outlets as far afield as *Australia Age* in Melbourne and the *New Statesman* in London; academics, including a medical professor at Harvard and the former commissioner of health for the City of New York; and—perhaps most impressively—two US government agencies, including the Federal Housing Administration, which included it in a fact sheet sent out to newspapers across the country.

Probably pride of place in the roll call of those republishing the admitted hoax goes to the *Boston Herald*, which printed the bathtub story as a stone-cold fact on June 13, 1926—a mere three weeks after they had published Mencken's "Melancholy Reflections" admitting it was fiction.

On and on it went, nothing seeming capable of removing this small shard of nonsense that had become lodged in the public psyche.

A few decades after its publication, it had made it all the way to the leader of the free world. In 1951, in a wide-ranging interview that President Harry S. Truman gave to the *New Yorker*'s John Hersey (which ran over five separate editions, following that magazine's long-standing tradition of believing that word limits are something for other, lesser outlets to care about), the president repeated the bit about Fillmore installing the first bathtub in the White House.[82]

The exchange that (as relayed by Hersey) occurred when an aide stepped in to correct the president is worth noting in full, because it's a pretty incredible baloney lasagna, bullshit sheet piled densely upon bullshit sheet.

"That's not true," the aide says. "That's from Mencken."

This debunking is not immediately accepted. "The President," Hersey writes with commendable understatement, "seemed reluctant to let go of his belief."

Truman insists that it *has* to be true, because he's "seen a paper the American Medical Association drew up claiming that the vapors from the tub were dangerous for the president's health."

That's the president of the United States of America, there, insisting that he's personally viewed a historical document that never existed because the thing it's about didn't happen.

"No," the aide responds, "I'm afraid Mencken invented that, too."

Which…he didn't, because that particular lie appears nowhere in Mencken's article. It's a completely original falsehood, sprung fresh from the brain of the world's most powerful man.

The president seems a little baffled by this revelation. "I could swear those AMA fellows didn't think it was a hoax about the tub," says the only person in history to order a nuclear attack, attempting to reconcile this new information with the tone of voice that he remembers some imaginary doctors using in a manuscript that his brain invented because a newspaper article from thirty-four years previously has made him believe a lie about baths so fiercely that his mind has had to overwrite reality to accommodate it.

"I feel robbed of a fact too, sir," the aide concludes, Smithers-like.

Now, of course, we have only Hersey's account of this conversation to go on, and obviously that could be every bit as inaccurate a recounting of fact as any other piece of journalism mentioned in this chapter. (If we're completely honest, some of the dialogue sounds a little stagey.) But, nonetheless, I choose to accept it—because, hey, it's the *New Yorker*, and at some point you've just got to trust somebody. Those motherfuckers have more fact-checkers than I have concerned texts about deadlines from my publisher.

Anyhoo, a year after the *New Yorker* interview was published, Truman gave a speech in Philadelphia, and he told the bathtub story again.[83] The whole malarkey of him being told that it was untrue, and then that conversation being published in one of the most prestigious outlets in the country so that everybody knew he believed an untrue thing…that apparently didn't do enough to shake his attachment to the tale.

And so, for decades to follow, the bathtub hoax continued to cheerfully reproduce itself down the generations. It even entered the twenty-first century alive and well: in both 2001 and 2004, the *Washington Post* published articles treating it as real history, before awkwardly having to print corrections.[84]

Some facts, apparently, are just so good that we can't allow them to not be true.

What does this all mean for the news industry and its noble pursuit of truth? Mencken may have put it as well as possible in a follow-up article he wrote in July 1926, inspired by the *Boston Herald* making a tit of itself. His words are a fairly bitter reflection of the business of journalism generally, but they also hit on the central point that falsehood has an inherent advantage over truth, simply because it's unconstrained by the tedious stranglehold of reality.

"What ails the truth is that it is mainly uncomfortable, and often dull," Mencken wrote. "The human mind seeks something more amusing, and more caressing. What the actual history of the bath tub may be I don't know: digging it out would be a dreadful job, and the result, after all that labor, would probably be a string of banalities.

"The fiction I concocted back in 1917 was at least better than that."[85]

4

THE LIE OF THE
LAND

There can be few more awe-inspiring sights in the world than the Mountains of Kong. The great mountain range bisects western Africa, its towering, snowcapped peaks first rising from the plains in western Senegal, before sweeping down through Mali and northernmost Guinea. In contrast to the surrounding landscape, the looming summits appear strikingly blue, a stunningly beautiful vista, albeit one that is barren and inhospitable. It's here that the rivers of the region—most notably the vast, winding Niger—find their origin, the frigid, surging meltwater pouring down from the vertiginous granite peaks in torrents between jagged ridges of quartz. Those rivers not only bring life-giving water to the plains below, they are laden with gold dust eroded from the mountaintops, carrying down with them a cargo that for centuries brought both prodigious wealth and much strife to the people who lived in Kong's shadow.

Onward the Kong range sweeps for more than a thousand miles, through Burkina Faso, past Ghana and Togo and Benin, and on into Nigeria, the low plains and gently rolling hills of those countries harshly split in two by what one nineteenth-century authority described as "towering masses of granite... upstanding outcrops resembling cathedrals and castellations in ruins; boulders of enormous dimensions; pyramids a thousand feet high, and solitary cones which rise like giant ninepins."[86] From there, it extends farther still, striking out, away from the southward-curving coast, deep into the interior. Eventually, the Mountains of Kong meet that other great mountain range of Africa, the eastern Mountains of the Moon (where the fountains that provide the source of the Nile are found), joining together into a single belt of impassable rock that bifurcates the entire continent, cutting off north from south.

Hmm. Right. Let's pause there for a second.

It's possible you may have some questions, at this juncture. Particularly if, let's say, you live in any of the aforementioned countries, or have visited them, or maybe just if you have a basic knowledge of geography. Those questions may include, but are not limited to, "Eh?" "You what?" and "There aren't any mountains there, fuck you talking about?"

To which I would simply say this: If there aren't any mountains there, then why do the Mountains of Kong appear on almost every map of Africa produced in the nineteenth century, and even into the twentieth century? Why are there multiple descriptions of their soaring granite edifices and inhospitable conditions, from European explorers who claim to have been there? Who are you going to believe: a handful of white dudes, or literally the entire population of the region?

I think we all know the answer to that one.

But the underlying question is a valid and fascinating one: Just how exactly did virtually every authority in Europe and America, for over a century, hold on to a steadfast belief in a massive mountain range that (to be perfectly clear) absolutely does not exist? I mean, they're *mountains*. There's not a huge amount of ambiguity in mountains. They're either there or they're not.

The answer to that question is still something of a mystery—but it's one that sheds light on exactly how we get things so wrong, so often.

Because, as this chapter will show, we haven't just spent history inventing wild falsehoods about events that happened *in* the world—we've done a pretty good job of inventing nonsense about the world itself. From imaginary mountains to wholly fictional countries and wildly improbable tales of far-off lands, the bullshitters of history have cheerfully exploited the fact that, traditionally, it's been quite hard to go and check when someone tells you about what it's like on the other side of the world.

Our history of geographical nonsense is, among other things, an example of effort barriers and information vacuums manifesting themselves on a grand scale. For most of humanity's time on this planet, long-distance travel was slow, hazardous and rare—many people would never travel any significant distance from their place of birth—and we didn't have the advantage of being able to hop on a plane or send up a satellite to take pictures of stuff from the air.

Given all this, it's pretty understandable that our conception of the planet on which we live might have been a little hazy. Mapmakers didn't really have a lot to go on and often had to improvise to fill in the blank spaces (even if the notion that they wrote "here be dragons" in the uncharted areas is, sadly, almost entirely a myth).[87]

But, even though the lack of information is a legitimate excuse, the ways in which this gap in our knowledge became filled with wild and ridiculous fictions tell us quite a lot about how bullshit spreads.

This much we know about the Mountains of Kong: the whole silly affair all started in 1798, with James Rennell's "A Map, shewing the Progress of Discovery & Improvement, in the Geography of North Africa."[88] Rennell is the first to plonk a huge mountain range in West Africa and call it the "Mountains of Kong." And, very quickly after that, everybody else picked up on this completely untrue notion and decided it *obviously* must be correct. For almost a century after that, most of the major maps of Africa produced in Europe and America featured the Mountains of Kong (over 80 percent, according to the definitive academic study of the mountain range's appearance and subsequent disappearance),[89] and numerous explorers reported back from the region, saying that they'd seen, or even climbed, the mountains, despite them not existing.

The curious thing about this is that Rennell wasn't just any old hack mapmaker, chucking around rivers and mountains for fun and saying, "That'll do," before trotting along to the pub. In fact, he was widely seen as just about the best in the cartography business at the time. One of his key skills was using his knowledge of the principles of geography and geology to make sense of the fragmentary—and often contradictory—reports that would come back from explorers. In fact, this might be what tripped him up.

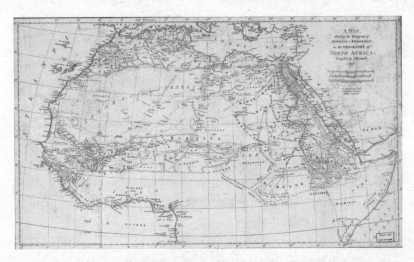

A detail from James Rennell's 1798 map, with the Mountains of Kong running across the exact place they aren't.

This was a time when not many Europeans had made it terribly far into the African continent—it was many decades before the Age of Empire's mad "scramble for Africa" would kick off—and quite a few of those who did make it had, if we're honest, very little idea what they were doing.

As a result, many maps from this era just sort of…shrug, when it comes to depicting most of the continent beyond North Africa and a few areas near the coast. Some leave central Africa pretty much empty, others sprinkle it with geographical features largely at random and yet others fill the space by drawing some pretty pictures of elephants.

Rennell's invention of the Mountains of Kong came because he paid too much attention to a throwaway comment from one of the few explorers to have visited the region, and then he filled in the rest, with unfortunate results. The explorer in question was Mungo Park—the dashing Scot whose diaries Richard Adams Locke would unsuccessfully try to fake, several decades later. At the time, everybody assumed that Park had died on an expedition to find the source of the Niger River, only for him to rock up unexpectedly in 1797, after several years off grid, and be all, "Guess who's back, bitches?" He hadn't found the source of the Niger, exactly, but he had at least traced it for several hundred miles along its somewhat eccentric route.

Park's stories of his journeys were quite the attention grabber, and Rennell was drafted in to provide illustrations for them. And it's here that the seed was planted, when Park said at one point, "Toward the south-east appeared some very distant mountains, which I had formerly seen from an eminence near Maraboo, where the people informed me that these mountains were situated in a large and powerful kingdom called Kong."[90]

Park was very likely telling the truth. He had been somewhere near Bamako in Mali at the time, and there was indeed a powerful kingdom called Kong not too far away (the Kong Empire, which had its capital in the modern-day Ivorian town of, that's right, Kong). And there are in fact some highland areas in the region that have occasional large rocky hills—"inselbergs," if you want to get geological—that you could maaaybe, depending on whether you're particularly flexible with your definition of "mountain," describe as mountains.

What there definitely *isn't* is a vast, impassable mountain range stretching for hundreds of miles. Which, you'll notice, is not something that Park actually says. He just says there were "some very distant mountains." It's not much to go on, and he doesn't really clarify much beyond that—which, in fairness, you can't really blame him for, on the grounds that, the day after he spotted the mountains, he was robbed by bandits who stole his horse, stripped him naked and left him in the middle of nowhere in the blazing midday sun. That's not really the kind of thing that puts you in a mountain-clarifying mood.

But Rennell latched on to Park's mention of some mountains in the kingdom of Kong and ran with it. He did this simply because it confirmed a pet geographic theory of his: the reason the Niger takes such a confusingly roundabout course through the region is because of the mountainous landscape. The Niger, you see, has rebelled against traditional river behavior, like flowing into the nearest sea, instead deciding to go on a loopy 2,600-mile inland jaunt that takes it right to the edge of the Sahara Desert, before it dramatically pulls a U-turn and heads on down to the Gulf of Guinea. This confused a lot of people for quite a long time. When Rennell made his map, pretty much all that was known about the Niger for sure was that it was very big, it presumably started somewhere, and then it kinda...went somewhere else.

Rennell's reasoning (which is honestly not that dumb) was that it must have its origins high in a lengthy mountain range that provided a physical barrier, thus directing its course eastward, away from the sea. He wasn't the first to think this—various huge mountains in the region had been hypothesized and added to maps throughout the 1600s, but had mostly fallen out of mapmaking fashion by the late 1700s.

And so he seized on Park's words, mixed them in with his theory, and argued, "They prove, by the courses of the great rivers, and from other notices, that a belt of mountains, which extends from west to east, occupies the parallels between ten and eleven degrees of north latitude."[91] Note that he doesn't go for "suggest" or "imply" or "make me reckon" here; he jumps right to "prove."

In other words, Rennell put two and two together and made a massive fucking mountain range.

And that would have been that, little more than an obscure footnote in the history of cartographic errors, if it wasn't for what happened next: everybody else immediately started copying Rennell, because he was a very good mapmaker and nobody wanted to look like the big stupid idiot who didn't even have the Mountains of Kong on his map.

The duplication of the error began almost immediately, when, in 1802, Aaron Arrowsmith published his new map. Not only did he copy the Mountains of Kong, but he also went one very large step further—he extended the mountain range across half of Africa, where he joined it up with the Mountains of the Moon to create the continent-spanning impasse that was mentioned at the beginning of this chapter (it's probably worth noting that every part of that rather florid opening description is taken directly from contemporary nineteenth-century descriptions of the imaginary mountains).

Arrowsmith's work is a fine example of the "sod it—just leave the middle blank" school of African maps; he would probably get credit for admitting the limits of his knowledge, if it wasn't for the fact that he'd just drawn a great slashing line of imaginary rock across the whole continent, ~~like an editor striking out an author's unhelpful simile~~.

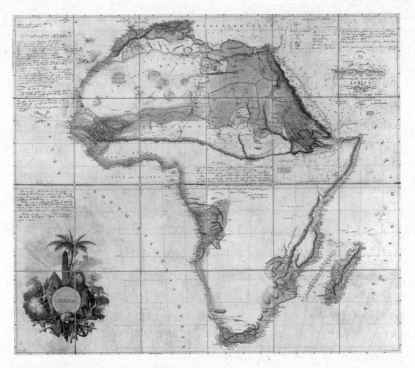

Aaron Arrowsmith's map, with the Mountains of Kong meeting the Mountains of the Moon, and a whole load of fuck-all in the middle of Africa.

Next up was John Cary, perhaps the only cartographer in Britain with a better reputation than Rennell, who also added the Mountains of Kong—and, in doing so, firmly planted the mountain range into the category of "stuff smart people know about." With both Cary and Rennell on Team Kong, it was virtually guaranteed that everybody else would follow suit. Like Arrowsmith before him, Cary joined up the Kong range with the Mountains of the Moon to create a continent-wide belt of impenetrable rock.

It's probably worth noting, at this juncture, that the Mountains of the Moon don't exist either.

These nonexistent mountains have an even longer history than the Mountains of Kong, with references to them as the supposed source of the Nile dating as far back as Ptolemy's *Geographia* in AD 150, and being repeated a thousand years later in the work of Arabic scholars such as Muhammad al-Idrisi. Unlike the Mountains of Kong, they're an almost permanent fixture on early maps of Africa, from the 1510s onward. It was only in the second half of the nineteenth century, when a pair of sickly, bickering Brits called John Hanning Speke and Richard Francis Burton managed to identify Lake Victoria as the true source of the White Nile, that Europeans finally came around to the idea that maybe the Mountains of the Moon didn't actually exist.[92]

Throughout the rest of the nineteenth century, most maps of Africa faithfully included the Mountains of Kong and (very often) the Mountains of the Moon. It's only toward the end of the century that some mapmakers started to get a little mountain-shy and began to wonder if maybe this vast edifice of rock might not be as massive and all-encompassing as common knowledge suggested. But, even then, doubters had to fight against a rather powerful source of evidence: the people who'd been there.

For decades, reports came back from explorers saying that they had seen or even crossed the Mountains of Kong. This one, honestly, is a bit of a puzzle, as they even included some people with generally sound reputations for not being complete flakes—such as Hugh Clapperton, a man of such fortitude that he once survived a grueling journey of 133 days with a traveling companion he hated so much they didn't speak once during that whole time.

The best explanation is probably also the simplest: these travelers came across some of the hills and inselbergs that dot the uplands of the area and simply assumed that they had survived the fearsome Mountains of Kong, because *that's what was on the map.*

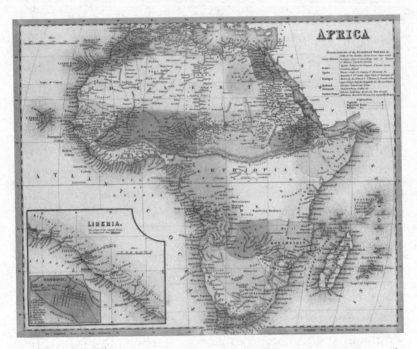

An 1849 map of Africa showing the Mountains of Kong and the Mountains of the Moon.

It was a perfect bullshit feedback loop of everybody assuming that everybody else must be right, adjusting their evidence to fit the theory, and then everybody else taking that as further evidence that the theory was correct all along. The explorers imagined the mountains because they were on the map; the map-makers took the explorers' words and fed them back into the next generation of maps. And so the fictional mountains endured.

Nowhere is this better demonstrated than in the paper that Captain R. F. Burton delivered at a meeting of the Royal Geographical Society on the evening of June 26, 1882.[93] Burton was a famed explorer and Orientalist, whom nature had endowed with a prodigious beard, a tongue for languages and a marked fondness for Eastern erotica. In his address to the Society, he stood up to stoutly defend the existence of the Mountains of Kong, which, he noted regretfully in opening, had "almost disappeared from maps" (not quite true—they were still a fixture on many maps produced at this point, even if the length of the range had cautiously retreated from some of the wilder assumptions in early maps). It's Burton we have to thank for the dramatic description of "towering masses of granite" mentioned earlier.

Burton makes the same key mistakes that Rennell made, just with another eight decades of evidence to misinterpret. He places way too much emphasis on theories about rivers, insisting "that such a chain must exist is proved by the conduct of the Gold Coast streams." He leans heavily on the reports of explorers like Clapperton and John Duncan (who is said to have "crossed the whole breadth of these Kong Mountains"), even when their actual words leave considerably more wiggle room than he suggests. For example, while Clapperton's editor added a chapter title that boasts of his journey "over the Kong mountains," Clapperton never actually uses that name; the actual description he provides mostly talks about "hills." The few mentions of mountains also give a height for those mountains of "six or seven hundred feet."[94] Which…doesn't really count as a mountain.

And, most notably, Burton casually dismisses the testimony of "a native guide" whom he'd spoken to himself on a visit to the area, who "knew the Kong village but not the Kong Mountains"—despite the fact that the village of Kong was supposedly located right at the foot of the mountains. Because, sure, why *would* you pay attention to somebody who actually lives in the area, going, "Oh, yeah, I know that village, but, nope, no idea about the huge mountain range that's right next to it"?

Burton wasn't alone in this. As Thomas Bassett and Philip Porter—the academics we have to thank for their study of Kong's history—note dryly, "There is evidence…that Europeans received first-hand accounts from Africans that the mountains did not exist in certain areas, but this information was usually ignored."[95]

In the end, it was left up to a French military officer named Louis-Gustave Binger to debunk the mountains. In 1888, he traveled to the area and was surprised to find "on the horizon, not even a ridge of hills! The Kong mountain chain, which stretches across all the maps, never existed except in the imaginations of a few poorly informed explorers."[96]

But, even after Binger dropped his "no mountains" bombshell, the Mountains of Kong enjoyed a lengthy afterlife—appearing in a few maps throughout the 1890s, one in 1905, and even turning up as late as 1928 in the *Oxford Advanced Atlas*, which clearly wasn't advanced enough to have updated its West Africa section in forty years. The mountain range may have passed away, but its ghost lived on.

While the nonexistent mountain ranges of Africa may be one of the more dramatic examples of the nonsense we've believed about the world around us, they're far from alone. History is littered with nonexistent lands and fictional places.

One of the longest-standing mythical lands is the "Kingdom of Prester John"—a supposedly utopian and vastly wealthy land, located vaguely in [waves hand airily] "the East." It was purportedly ruled by a Christian monarch named—that's right—Prester John. The figure of Prester John had been confined to folklore until, in the twelfth century, a hoax letter claiming to be from Prester John himself (pledging his assistance in the Crusades) began circulating.

Both the author and the purpose of the hoax are sadly lost to the mists of time, but the letter kick-started a persistent and long-running belief in the existence of this sparkling unicorn of an empire. For five centuries, explorers set out to discover this great lost land, mapmakers put it on their maps and real-life figures were said to be descendants of Prester John.

This was despite the fact that nobody was ever entirely clear on where it was supposed to be. Over the course of a few hundred years (as fashions changed and parts of the map became filled in with lands that definitely *weren't* it), the supposed location of Prester John's kingdom went on a globetrotting adventure, starting in Asia, before traveling to various parts of Africa and finally settling somewhere in Ethiopia, conveniently close to the nonexistent Mountains of the Moon. It wasn't until the seventeenth century that people began to reluctantly accept that it had never existed.

It was the tales sent back from the early European explorers that helped to fill in the holes in maps, but their testimony was often quite a way short of reliable. For example, the land of Patagonia was, according to a description of Magellan's voyages in 1520, populated by a race of giants. The notion of the Patagonian giants became a mainstay of European explorers' tales for many years, with descriptions of their height fluctuating over time: the giants were twelve feet tall at one point in the late sixteenth century, which was downgraded to nine feet tall by the eighteenth century.

The reality is that the "giants" were most likely the indigenous Aónikenk people, who are not in fact giants—they're merely what could be described as "quite tall," in that they are often around six feet. The idea that these people were "giants" can mostly be put down to the fact that, at the time, most Europeans were short-asses.

It's not just false claims about what you saw on your travels; sometimes it's the journey itself that's fictional. History is littered with people who claimed to have undertaken voyages or treks that they never did. To take one example—purely because it's on brand for this book—for many years biographers of Benjamin Franklin were fooled by a French author called St. Jean de Crèvecoeur, who claimed to have sailed with Franklin to Lancaster, Pennsylvania, in 1787 to attend the founding ceremony of Franklin College. In reality, neither man was there: de Crèvecoeur simply made it up because he wanted his countrymen to be impressed by his friendship with their favorite American.[97]

In the annals of unreliable explorers, the business of discovering islands provided the most fruitful territory, as it was extremely easy to claim that you'd found one in the middle of the ocean and it would be very hard for anyone to check. Also, you got to name the island after either yourself or (often) a rich benefactor, which was extremely good for business.

Perhaps the most enthusiastic proponent of fake island discovery was a gentleman named Benjamin Morrell, a nineteenth-century adventurer and profoundly dedicated bullshitter, born in the still-young nation of the USA in 1795. As Edward Brooke-Hitching notes in his excellent book *The Phantom Atlas* (a compendium of "speculative geography," which I can heartily recommend if you're a fan of the things in this chapter), Morrell came to be known as "the biggest liar of the Pacific"[98] thanks to the wildly untrue tales of discovery he published in his book *A Narrative of Four Voyages*. Among his "discoveries" were New South Greenland (which doesn't exist), Byer's Island (doesn't exist, named after a wealthy bloke he wanted to impress) and Morrell's Island (named after himself, also doesn't exist). Despite none of them existing, several remained on navigational charts for over a century, which can't have been terribly helpful for people trying to navigate.

It would be easy to assume that such fictional geography is now a thing of the past—understandable in the context of the lack of information available to those living in the past, but not relevant to us living in the age of satellite photos and Google Maps. Easy, but not necessarily accurate, because many of the misconceptions of the past continue to echo today. Indeed, at least one of them actually found its way onto Google Maps itself: a remote patch of land a few thousand miles from Australia, named Sandy Island, which appeared on maps for well over a century until, in 2012, an Aussie ship happened to sail past where it was supposed to be and confirmed that, not only was there not an island there, but the seabed at no point came within half a mile of the surface. Google, along with other organizations, including the National Geographic Society, hurriedly deleted it from their maps.

The persistence of unreal lands is perhaps not surprising, given that land is something people really, *really* want. Land gives you a home, yes, but—possibly more important—land can make you very rich. You can see that most clearly, perhaps, in the tale of the nonexistent island of Bermeja, a spurious landmass supposedly in the Mexican Gulf, off the north coast of Yucatán. First appearing on maps in the sixteenth century, it had mostly vanished from charts by the early twentieth century, only to undergo a dramatic revival of fortunes when the Mexican government realized that, if it existed, it would enable them to claim a really, really large patch of the Gulf's lucrative oil fields as their own.[99] For years, Mexican ships hunted fruitlessly for the fictional island—and, while they've had to admit finally that there definitely isn't an island there, to this day there are many who insist that there must have been an island there at *some* point. Some Mexican legislators even accused the CIA of making the island vanish.

So often, our wrong beliefs about the lands we live on stem from our old friend, motivated reasoning: simply, we want it to be true, because land means wealth and power and—for many individuals—glory. Nowhere is that seen more clearly than in the tale of the battle to be the first explorer to reach the North Pole. Or, rather, to be slightly more precise, the battle to get the *credit* for first reaching the North Pole.

This battle was launched in the pages of the American press in 1909. On September 7, the *New York Times* announced triumphantly on its front page, "Peary Discovers the North Pole after Eight Trials in 23 Years," awarding the victory to Robert E. Peary. That sense of triumphalism was in no way dimmed by the fact that, just five days earlier, the *New York Herald*'s front page had declared, "The North Pole Is Discovered by Dr. Frederick A. Cook."

The *Herald*'s story must have come as a bit of a blow to Peary, who had just returned from his expedition, eagerly anticipating the opportunity to announce his remarkable achievement to the world—only for his former friend and shipmate, Cook, who had been missing (and generally presumed dead) for the past year, to suddenly reappear and claim that he'd actually made it to the Pole in 1908. In the *Times* article, Peary furiously denounced Cook as a fraud.

The fight for credit played out initially in the court of public opinion, and, to begin with, Dr. Cook seemed to have benefitted from getting his story out first. He was greeted by rapturous crowds when he landed in New York. Newspapers across the country polled their readers on whose claim they believed, and Cook repeatedly came out the overwhelming winner.

But Peary was a savvy PR operator and quickly called in favors in his campaign to have Cook's claim discredited (he also, it was claimed by the *Herald* in an effort to protect its scoop, bribed at least one witness). Traveling light on his return journey, Cook had left many of the records of his expedition behind with an acquaintance in Greenland, who promised to deliver them to New York. Unfortunately for Cook, the ship that his friend took on his journey back to New York was none other than Peary's—who, in a supremely petty move, refused to let him bring any of Cook's belongings on board.

After getting the news that the evidence he believed would vindicate him was not turning up, Cook began to sink into depression, and, a few months later, left the US for Europe, where he would spend a year in exile, writing a book. The *New York Times* crowed about the news of his "disappearance," branding his tale "the Greatest Humbug of History" and "the most astonishing imposture since the human race came on earth."[100]

(It's possible that the *New York Times*'s decision to unhesitatingly support Peary's claim might have had *something* to do with the fact that it had already paid $4,000 for the rights to cover his expedition.)

Backed by a powerful newspaper, the National Geographic Society (who had sponsored Peary's expedition in the first place) and large parts of the establishment, Peary's version soon overtook Cook's and eventually came to be the commonly accepted story. His journey to the Pole was officially recognized by Congress, and, having been a civil engineer for the US Navy, he was promoted in retirement to the position of rear admiral and awarded a pension of thousands of dollars a year.

In the century since, supporters of both Cook and Peary have furiously debated whose claim was legitimate. Was Cook truly a fraud who had tried to deny Peary his rightful triumph, or was Peary a sore loser who pulled strings to unfairly steal the honor of the man who beat him?

The truth, you'll be delighted to learn, is "none of the above." In fact, it turns out they were both lying.

Today, the expert consensus is that, in all likelihood, neither man had made it to within a hundred miles of the Pole—and they both fabricated evidence to cover up their failure.

Cook's claim is the easiest to demolish. For starters, he was already embroiled in a scandal over his honesty—one which Peary cynically, but accurately, latched on to in his PR campaign. His claim to have been the first to scale Denali (North America's highest mountain, known at the time as Mount McKinley) had been widely called into question. That skepticism only grew after it emerged that Cook had cropped a photograph of his companion triumphantly standing on the "summit," to remove what was very obviously a much higher peak in the background. Much like you on Instagram editing out the McDonald's in the corner of your idyllic yoga retreat.

Any remaining doubt about this vanished a year later, in 1910, when another expedition tried retracing Cook's steps and discovered the peak where the photo had been taken—which turned out to be about twenty miles away and fifteen thousand feet lower than the true summit. (Rather pleasingly, this outcrop is now officially known as "Fake Peak.")

The ol' picture-swap trick turned out to be a favorite tactic of Cook's, as it was subsequently discovered that pictures he had submitted of the "North Pole" were actually old photos he'd taken in Alaska. The diary he offered as evidence of his travels had clearly been written after the fact; his Inuit guides later said that they hadn't made it to the Pole and an island that he claimed to have discovered en route was subsequently shown to, er, not exist. The final nail in the coffin of Cook's reputation came in 1923, when—having got out of the exploring game and gone into the oil industry—he was convicted and jailed for mail fraud. Which pretty much guaranteed that he'd go down in history as a dodgy sort.

This left the field wide-open for Peary to be crowned the true discoverer of the Pole. So comprehensively had Cook's reputation been trashed, and so public had been the feud between them, that Peary seems to have had a pretty easy ride—presumably on the flawed assumption that, if one of them was lying, then the other *must* be telling the truth. As such, his story came to be widely accepted for most of the twentieth century.

This is a bit odd, because, even at the time, there had been serious doubts about his honesty, and, in retrospect, quite a lot of his behavior should have thrown up red flags. There was his diary, which he presented to Congress as proof—a diary that the congressmen couldn't help but notice was surprisingly pristine for an item that had supposedly been written daily by someone whose hands were covered in grease, in a hostile environment, where it was impossible to wash. There was also the fact that Peary refused to let anybody else examine his records. And there was the fact that he, too, claimed to have discovered an island en route, which subsequently turned out to not exist.

But the biggest red flags came from the implausibility of his own description of his journey. The thing about the North Pole is that it isn't on land—it's covered by sheets of ice, and the thing about sheets of ice is that they drift around quite a lot. There's also not much in the way of landmarks to navigate by in the vast, empty white expanse. To walk to the Pole, you need to take regular navigational sightings to check both that you haven't wondered off course, and that the ground beneath your feet hasn't shifted away from where you expected to be. This is something that Peary certainly *could* have done—unlike Cook, he was an expert navigator, as were several other members of his party—but, the weird thing is, he didn't take any sightings at all on the entire trek toward his goal.

And yet, according to his own description of his journey, he had managed to walk nearly five hundred miles across the shifting, landmark-free ice in an arrow-straight line, directly to the Pole. Not only that, but, a week before he supposedly made it to the Pole, he sent most of his party back, including all of the other trained navigators. Which is a bit suspicious, but not as suspicious as how, immediately after they were sent away, the supposed pace of his travels mysteriously doubled to an implausible seventy-one miles a day—a remarkable feat for someone who had lost most of his toes to frostbite on an earlier expedition. (The doctor who had saved his remaining toes was...his future nemesis, Frederick Cook.)

In the end, Peary took one single navigational sighting, after which—according to his traveling companion, Matthew Henson—he returned looking thoroughly miserable, and refused to tell anybody what the result was. But, the next day, he simply declared that they were at the Pole, stuck a strip of the American flag in a tin and buried it under the ice. Then they went home.

While the debate about which of the two rivals was the true polar pioneer would rage for decades afterward, the consensus opinion now is that neither of them made it to the North Pole:

Cook never came close, and Peary probably missed his target by somewhere between sixty and one hundred miles. In reality, the first expedition across the ice sheets to reach the Pole didn't happen until 1968, and even then they did it on snowmobiles.

It's perhaps worth noting here that Cook and Peary were far from the only Arctic explorers to have historical questions hanging over them: take, for example, our old friend Vilhjalmur Stefansson, the future cataloguer of bathtub hoaxes. In 1913 he led a Canadian expedition to explore for new lands between Alaska and the Pole, only for his ship to become trapped in the ice, in danger of being crushed.

Stefansson promptly announced that he was going ashore to hunt for food; while he was gone the ship was carried away by the ice and eventually sank. Eleven of Stefansson's crew died before they were rescued, while Stefansson himself continued to cheerfully explore the Arctic on sled for another four years, not seeming terribly concerned about what had happened to his ship or crewmates. At least some of the ship's survivors later said that they suspected Stefansson had deliberately abandoned them.)

While Cook and Peary both resorted to deceit in order to cover up a failure to achieve their goals, other geographical bullshitters throughout history have done it the other way around: setting themselves up for failure by spinning tall tales, which reality could never hope to match.

That's what happened to a man named Lewis Lasseter, who, in 1930, led a party into the central Australian desert in search of wealth beyond imagination. They were looking for an enormous "reef" in the middle of the outback, one that was made of gold and would make them all richer than kings.

The obvious point, here, is that there isn't a nine-mile-long gold reef in the middle of Australia. Lasseter claimed that he'd seen it with his own eyes, having stumbled upon it while lost in the desert in either 1897, 1900 or 1911 (the year varied with the telling). He had then been unable to retrace his steps, the story went, but he'd spent several decades trying to raise the funds to mount an expedition to rediscover it.

It's still a matter of debate as to whether Lasseter was simply mistaken, actively delusional or an outright scammer. Possibly a little from each column, as is often the way. But, nonetheless, as the Great Depression hit worldwide, his tale convinced a powerful union boss to give him backing on the off chance he was right. So a party of eight explorers set off, well equipped with a plane and some trucks, looking for the gold.

It quickly became apparent to his companions that Lasseter had absolutely no idea where he was going and, furthermore, had clearly never even been in the bush before. The group searched around to no purpose, their trucks got stuck in the sand and their plane crashed, hospitalizing the pilot.

One by one, they came to the conclusion that Lasseter was full of shit and abandoned the expedition—eventually leaving Lasseter with only a dingo shooter called Paul and some camels as his companions. Lasseter soon told Paul that he'd found the reef but refused to say where. After a brief fistfight, Paul left, too, and then finally (according to his diary) Lasseter's camels ran away from him while he was taking a poo.

Lasseter's remains, along with the diary, were found in the desert the next year.

Despite the fact that there definitely isn't a reef of gold in the middle of Australia, numerous further search parties went looking for it in the subsequent decades, and to this day, every few years, somebody new will claim to have discovered the location of Lasseter's Reef, which, to reiterate, doesn't exist.

Lasseter may have genuinely believed in his reef; it's hard to explain why, if it had simply been a scam, he persisted in his search long after everybody else had given up. Like many of those who spun false tales of nonexistent lands, he could simply have been mistaken and then doubled down on his false belief, due to enthusiasm or shame or simple confirmation bias.

In that, he's hardly alone. Indeed, one of the most incredible tales of geographical nonsense in history belongs to a man who had no excuse at all for the untruth of his claims and yet persistently acted as though they were certainties. But to tell that story, we'll need to delve into the murky world of con artists, scammers and bunco men, where we'll meet perhaps the greatest grifter of all time—the man who defrauded a country by inventing a whole other country.

5

THE SCAM
MANIFESTO

When the first settlers arrived off the shore of the territory of Poyais in February 1823, they already had a very clear idea of what to expect from their new home. As the *Honduras Packet* dropped its anchor just outside the Black River lagoon, the colonists aboard must have been eagerly anticipating the wealth that awaited them in their new life. The land of Poyais was both beautiful and fertile, they knew. Its warm Central American climate was a far cry from the wintry London they'd left, two months earlier, and was said to do wonders for the health. Its rich soils could grow three crops a year, promising any enterprising farmer a ready fortune for minimal effort. The country's long, winding rivers ran full of gold, small nuggets of it obtainable to any passerby simply by sifting the fine sands. Inside that lagoon lay the mouth of the Black River, home to the country's major trading port, and but a few miles distant was the capital city of Saint Joseph—a small but growing metropolis of fifteen hundred souls, with elegant architecture after the European fashion.

No boat yet having emerged from the lagoon to greet their new fellow countrymen, Captain Hedgcock fired one of the ship's cannons to alert the Poyers (as the citizens of Poyais were known) to their arrival. They waited excitedly for the harbor's representatives to row out and receive them.

And waited. And then waited some more.

Hmm. Still no boat.

The boat never emerged because, as the settlers discovered when they finally decided to go ashore by themselves, inside the lagoon there was no bustling trading port. When they searched for the capital, Saint Joseph, venturing miles up the river and cutting their way through the dense jungle, they didn't find a cosmopolitan city with wide boulevards, a bank and an opera house. Instead, they found some rubble and a few ruined huts that had been abandoned the previous century. Assuming that they must have landed at the wrong place, they checked against a detailed map of the nation, one which they had been given by the Cazique of Poyais himself: the war hero, descendant of nobility and inspirational ruler of this young country, General Sir Gregor MacGregor.

Nope. They were definitely in the right place.

The thing that the settlers hadn't quite deduced, at this stage—but which, perhaps, some were already feeling the initial stirrings of, down in that bit of the stomach that first realizes when you've made a terrible, terrible mistake—was that the reason there were no boats, and the reason there was no port and no capital city, wasn't because they'd been given bad directions.

It was, of course, because there's no such country as Poyais. This entire young nation existed almost exclusively in the mind of Gregor MacGregor, a man who somehow leveraged his fictional dominion to raise a fortune from investors in London, and who convinced hundreds of his fellow Scots to sell their worldly possessions, abandon their homes and cross an ocean (all while paying him handsomely for the privilege of a new life).

Within a year, most of them would be dead.

Some con artists work their grift by inventing imaginary businesses, or sick relatives or mysterious fortunes that can only be recovered with the assistance of a randomly emailed stranger. These people are like little babies compared to MacGregor, who invented an entire country.

We're kind of obsessed with con artists, swindlers, grifters and scammers of all sorts. Whether we see them as cruel exploiters of the vulnerable and gullible, or as having a perverse folk-hero status—turning unfair systems back on themselves—we can't get enough of them. That could be because we enjoy the schadenfreude of seeing others getting fooled, or because we revel in the paranoid fear that we could get tricked ourselves. Or it could be because of the way they seem to confirm what many of us secretly think about the social structures that separate the haves from the have-nots—that they're all phony, hollow facades that any of us could break down, if only we had the chutzpah to pretend to be something we're not.

MacGregor is now remembered by history as having orchestrated what the *Economist* once described as "the greatest confidence trick of all time."[101] But what's fascinating about him is that, to this day, it's still not entirely clear quite how much was real, how much was a very deliberate con and how much was simply self-delusion on the grandest of scales.

Ambitious, charismatic and at least provisionally charming, MacGregor was a man who felt very profoundly that he was destined for greatness. What's more, he consistently managed to maneuver himself tantalizingly close to that greatness...only to promptly drive his career off a cliff every time. To put it bluntly, if MacGregor had only put the same amount of effort into *actually* achieving things as he did into *pretending* that he'd achieved them, then he (not to mention several hundred impoverished or dead settlers) would have been a lot better off.

It's possible to understand why the settlers and investors who fell for MacGregor's scheme did so. He boasted a truly impressive pedigree: a Scottish noble who was a veteran of the British army, he had served with the legendary Fifty-Seventh Regiment of Foot—the "Die Hards"—at the Battle of Albuera. He had also fought in the Portuguese army and had been made a knight of the Order of Christ by Portugal in return for his service. Then, as many British military men of the time did, he traveled to Latin America to fight in their wars of liberation against the Spanish empire, becoming a general in the Venezuelan army and a hero to its people. Perhaps all this wasn't surprising—after all, he was the chief of the Clan Gregor, a descendant of the legendary Rob Roy himself.

Then ads began to appear in the press in late 1821 for the opportunity to buy land in the country of Poyais—for the marked-down early bird price of one shilling per acre, if you got in quickly, although the ads warned that the price would be going up in the coming months.[102] It must have seemed too good an opportunity for many to ignore. As the centuries-old Spanish empire in the New World finally crumbled, British eyes were hungrily looking at a new opportunity, and investment in Latin America was the hot new thing. By the summer of 1822, the ads weren't just for the chance to invest in land: they were advertising for settlers to go and start a new life in Poyais, as passengers on the "very commodious and comfortable" *Honduras Packet* (as one ad in *The Times* put it).[103]

To back this up, MacGregor launched a full-blown publicity campaign. He gave interviews in newspapers, he pressed the flesh in high society and he established offices for his imaginary country in London and Edinburgh. Not only that, but, at some expense, he even had an entire leather-bound book published, titled *Sketch of the Mosquito Shore*, supposedly written

Gregor MacGregor, looking fancy, as he appeared in Sketch of the Mosquito Shore.

by a "Thomas Strangeways, K. G. C.," who was described as "Captain 1st Native Poyer Regiment, and aide-de-camp to his Highness Gregor, Cazique of Poyais."

Sketch of the Mosquito Shore included a very nice portrait of Gregor looking regal on the first page, plus an idyllic drawing of the Black River lagoon filled with ships. It was this book that promised those rivers that would furnish settlers with "globules of pure gold," soils that could grow three crops per year, plus friendly native laborers who had a deep and abiding love of the

British and who would cheerfully work all year round in return for a small sum of money, or possibly just for clothing.[104]

(The book also took pains to point out, right at the beginning, that the Mosquito Shore isn't called that because it has loads of mosquitoes—ha, the very idea!—but rather because of the many small islands that dot its shoreline. Neither of these explanations is actually true; in reality, the Mosquito Coast is named, or misnamed, after the indigenous Miskito people.[105] It does have really quite a lot of mosquitoes, though, as the settlers would soon discover.)

In truth, much of the book was simply copied wholesale from several other books about the region that were decades out of date; the material that wasn't copied was pure fantasy, which testimony at a later libel trial would reveal was actually written by MacGregor himself.

But MacGregor went even further than the book to sell Poyais not merely as an opportunity to create a new colony but as an already-established country with a functioning government, extensive civic infrastructure and a vibrant culture. He would show people a copy of the "Proclamation to the Inhabitants of the Territory of Poyais," marking the official birth of his nation—a document which he had supposedly distributed across his lands before departing for London—declaring that the king of the Mosquito Shore had granted him the rights to the territory of Poyais in perpetuity. He invented a flag for the country, and a chivalric honor system, awarding potential allies in his scheme the "Order of the Green Cross." He had "Poyais dollars" printed and provided a chest of them to the settlers to set them up in their new home. He discussed the tripartite structure of Poyais's system of government. He persuaded an impressionable young Glaswegian clerk named Andrew Picken, who had dreams of the literary life, to write a poem and a ballad singing the praises of Poyais, with the impression given that these were the products of Poyais's own culture.

Picken would go on to become one of the foremost voices in telling other settlers the good news of the life awaiting them in Saint Joseph, following a wine-fueled chat during which MacGregor strongly suggested that he might be made the head of the national theater of Poyais. Many settlers were similarly promised major posts: one would be the lieutenant governor of Saint Joseph, another the head of the Bank of Poyais. One Edinburgh cobbler named John Hellie (or Heely, documents differ) sold his possessions and left his family behind after he was promised that he would become the official shoemaker to the princess of Poyais.

A Poyais dollar bill, as printed by Gregor MacGregor in Scotland, because there was no such thing as the Bank of Poyais.

Of course, what the settlers found when they got there was… slightly less impressive. They came ashore on the northern coast of modern Honduras, at the western end of the evocatively named Gracias a Dios (Thank God) region. The Black River is today known as the Rio Sico; the lagoon is known variously as Laguna de Ibans or Laguna Ebano.[106] It's still pretty remote and not many people live there, although there is an airport—well, a grass landing strip—and, according to Lonely Planet, a rather nice "eco lodge" catering to the tourist trade.[107] All told, probably a fair bit more hospitable than what the passengers on the *Honduras Packet* faced.

What they faced was nothing very much, other than a lot of jungle, some rubble and an American hermit living in a hut. There was no city, no town, no port and no trade. The rivers were notable for their lack of globules of pure gold. The first batch of settlers from the *Honduras Packet* spent some weeks obliviously trying to work out how they'd gone wrong and waiting for the Poyer authorities to contact them, all the while living on the shore in tents and rudimentary shelters.

Some weeks later, in March, a second colony ship—the *Kennersley Castle*—arrived, which was when things started to go even more deeply wrong. For a start, it raised the number of settlers from around seventy to over two hundred, which was a lot more mouths to feed and a lot more bodies to fall ill. Worse, there was immediately friction between the two groups. The new arrivals—who had been hearing Picken's tales of how wonderful Saint Joseph was all the way over—were particularly unhappy with what they actually found when they arrived. But they also couldn't understand why the man in charge, Colonel Hector Hall (the one who was supposed to take up the post of lieutenant governor), hadn't got on with the business of building more permanent shelter, or indeed why he wasn't even there to greet them.

Partly this was because the people chosen as colonists didn't really have the skill set you'd want if you were trying to build a town from scratch. They included the banker, some civil servants, a jeweler, a printer, several gardeners, a "gentleman's servant" and a number of cabinetmakers: all excellent professions, if there's already a bustling metropolis waiting for you, but of varying use if you need to construct anything bigger than a cabinet first.

But the main reason was that Colonel Hall had already realized what most of the colonists hadn't yet—that they'd been had, on an epic scale. No Poyer authorities were ever going to contact them. Leaving the shore and moving inland would have meant death. Building a settlement was pointless. With the *Honduras Packet* having sailed away during a storm, carrying many of their supplies with it, their priority now was rescue. Which was where he was: on an expedition to try to locate the vanished *Honduras Packet* and to make contact with King George Frederic Augustus (the nominal Miskito ruler of the region, who had been installed largely as a puppet of the British and was the one who had supposedly granted MacGregor the rights to Poyais in the first place).

On his return, having been informed that the king had very little idea what he was talking about, Hall was particularly unhappy to find that the new arrivals had let the *Kennersley Castle* sail away too. The settlers, for their part, were unhappy that he'd brought only one small barrel of rum back with him.

Things went downhill rapidly. Morale plummeted amid infighting, efforts to construct more shelter failed and, worst of all, with the coming of the rainy season—and the attendant mosquitoes, which turned out to be plentiful—the colonists began to get sick and die. Hall continued to keep his knowledge that they'd been conned to himself, fearing the reaction if word got out, but this only had the effect of increasing the distrust between the groups, especially when he kept on vanishing for long periods of time on mysterious missions. Despairing of ever seeing his family again, the poor shoemaker, John Hellie, shot himself dead in his hammock.[108]

Eventually, in May, after they had endured several torrid months, a ship from Belize discovered their miserable encampment. This brought good news and bad news. The bad news was that the country they thought they had emigrated to didn't exist; the good news was that they could leave it. That this was the best course of action was confirmed a few days later, when Hall returned from his latest expedition with a message from

King George Frederic Augustus: any land grant MacGregor had claimed was null and void, and the colonists were, in fact, illegally trespassing.

And so the broken colonists were ferried on a series of horrible, cramped voyages to Belize. Some were too sick to even make the journey; many more became sick or worsened on the voyage itself. More than half of them died: of around 270 settlers who had made the journey on the first two ships, perhaps only fifty ever made it back to the UK.

At which point, it's probably worth noting that this whole business—"shiploads of colonists depart Scotland for dreams of a new paradise in Central America, only for the dream to end in financial ruin, disease and death"—might sound a little... well, *familiar.*

That's because, incredibly, this wasn't the first time it had happened. A hundred and twenty-five years earlier, almost the exact same fate had awaited several thousand Scots, who, convinced by a smooth-talking salesman, had left to found a new colony in Darien, on the Isthmus of Panama. That time, it wasn't exactly fraud, just a wildly overambitious attempt to found a Scottish empire and prove some theories about global trade, but the effect was much the same. Around half the colonists wound up dead, and many investors were ruined. The whole affair was a profound national humiliation for Scotland, shattering its economy and helping to drive it into an eventual union with England.[1]

This does rather raise the question: How the hell do people fall for exactly the same thing just over a century later?

To answer that, we need to look at MacGregor himself and work out just how he was able to convince so many people that a country which doesn't exist was real enough that they should screw their lives up for it. This is where we hit a slight snag, be-

---

1   If you fancy reading more about the Darien affair, there's a hefty section on it in the author's previous book, *Humans: A Brief History of How We F*cked It All Up.*

cause, unfortunately for both MacGregor's reputation and our efforts at a balanced portrait of the man, he's at something of a disadvantage on the historical legacy front. Quite simply, an awful lot of what was written about him at the time was written by people who very clearly hated him.

MacGregor was extremely skilled at sweet-talking people and getting them on his side. Sadly for him, he was absolutely lousy at keeping them there.

First up, it probably won't come as a huge surprise to you that not only was MacGregor's country fictitious, but so, too, were large swathes of his biography. True, he was a MacGregor of the Clan Gregor, but he certainly wasn't the chief of it and wasn't a direct descendant of Rob Roy; he was from the less-favored side of the family. Yes, he'd served in the British army, but he wasn't one of the "Die Hards" of the Battle of Albuera, because he'd been quietly booted out of that regiment over a year before the battle, following what was euphemistically described as "a misunderstanding with a superior officer."[109] And it was correct that he'd subsequently been seconded to the Portuguese army...but he lasted only a few months before exactly the same thing happened again.

Precisely where his supposed Portuguese knighthood could have come from, in the space of a few months that was spent mostly pissing off his superiors, is left as an exercise for the reader.

MacGregor's problem was that, while he undeniably had some talents, his taste for the trappings of status far outstripped his diligence in acquiring that status. As the historian Matthew Brown (who is generally more sympathetic to MacGregor than most other writers) puts it, MacGregor was "a pretentious and status-obsessed man."[110] He'd married into a wealthy and well-connected military family and had followed the time-honored tradition of buying himself promotion through the ranks of the army, becoming more and more insufferable with every extra stripe. After he was kicked out of the service in 1810, rather than engaging in a period of self-examination, he seems to have doubled down, awarding himself the title of colonel and parading around Edinburgh with his wife, dressed in their finest regalia. As a particularly brutal biography of him, published in 1820 by one of his many enemies, put it, he enjoyed his freedom "with little foresight and less reflection."[111]

This all came to a sudden and tragic halt in 1811, when MacGregor's wife, Maria, died. Cut off from her family's wealth, he couldn't fund his fancy-pants lifestyle anymore. Cash-strapped and purposeless, he did what many other British ex-military types did at the time: went off to Latin America to fight against the Spanish. Specifically, he hightailed it for Venezuela.

It was in Venezuela that MacGregor came the closest to bridging the gap between his self-image and his actual accomplishments. Like many a gap-year student, he went off on a foreign jaunt and…found himself. He quickly became a close confidant of the great revolutionary general Francisco de Miranda, whom MacGregor regarded with googly-eyed admiration. An inveterate pleasure-seeker and legendary playboy, Miranda shared MacGregor's taste for the perks of power; unlike MacGregor, he was also a military genius. Not only did MacGregor get in with Miranda, but for a second time he married a woman with connections: Señora Doña Josefa Antonia Andrea Aristeguieta y Lovera, cousin of the now-legendary liberationist Simón Bolívar.

MacGregor's military record in Venezuela wasn't perfect, but all told it was pretty solid and included at least one achievement that he would be justly feted for. In fact, it could probably have been much better if he hadn't had the misfortune to arrive at a low point for the Venezuelan independence forces. He suffered some defeats but was hailed as a hero for commanding a vital month-long retreat from Ocumare in 1816, which saw him lead a force largely composed of recently liberated slaves as they valiantly fought off pursuing enemies—an exceptional rearguard action that allowed the pro-independence army to regroup. Finally, finally, MacGregor had the acclaim he craved—not from invented titles or throwing money around, but from hard-earned achievements.

Anyway, shortly after that, he seems to have had some kind of catastrophic falling-out with the Venezuelans and left their service. Oh, Gregor.

From this point onward, his career got increasingly wild, as he started to go freelance. In 1817, he tried to invade Florida and seize it from the Spanish. His troops ended up spending six months stuck on a small island that they had captured, before Gregor bolted and left them there. He led a disastrous attack on the Isthmus of Darien—the very place where Scotland had suffered such humiliation in the previous century. MacGregor explicitly referenced this in his efforts to raise troops, claiming that one of his ancestors had been on that ill-fated expedition, and painting it as an opportunity to redeem the nation's honor. It didn't end up that way; in Portobelo, MacGregor was caught, literally sleeping, by Spanish forces, and ended up having to jump out of his bedroom window minus his trousers and swim for safety (he couldn't swim).

It was this period that led to many of the unflattering portrayals of him. Michael Rafter, who served under him in Darien and whose brother was executed after the Spanish retook Portobelo, became determined to expose MacGregor—it's his biography that we've already quoted from, which sums our hero up by saying that "M'Gregor was spoiled by prosperity, and his versatility and haughtiness of disposition soon overturned his flattering prospects," which honestly seems fair.[112] Another account of the misadventure describes MacGregor commanding an action from a ship's quarterdeck with a glass of wine in his hand.[113] The *Jamaica Gazette* wrote it up with the less-than-flattering words, "He sets out upon a freebooting expedition against, as his partisans would call them, his enemies, and closes his career...by plundering his friends... The cause of this said great Leader is now, it seems, become completely useless, and the Hero himself completely unworthy of any more notice."[114]

Throughout this period, MacGregor was honing three traits that, in retrospect, seem like trial runs for the Poyais scheme. He had a talent for recruiting his countrymen to his missions, convincing scores of soldiers to leave Scotland and join him in adventures across the ocean. He was already building his fantasy honor system: the "Order of the Green Cross" got its first tryout in Florida. And he was giving free rein to his love of issuing bullshit proclamations and awarding himself the grandest of titles—after one expedition, Rafter writes, in "a lasting monument of the singular aberration of the human intellect, he had the unparalleled effrontery to style himself the 'Inca of New Grenada'!!"[115]

The thing is, though, none of this behavior was *especially* weird in the context of the Caribbean and Latin America at the time.[116]

I mean, okay, it was *kind* of weird, but it wasn't completely outrageous. Giving yourself fancy titles was close to standard operating practice for the caudillos of the region; with empires falling, rising and generally dancing the mambo, territory was in an almost permanent state of flux and always up for grabs; and speculative investments in Latin America were ten a penny on the London exchange, fueling a bubble that would burst not long after the few Poyais survivors straggled home.

And MacGregor does genuinely seem to have been granted some land by King George Frederic Augustus (who would often dispense land in return for political favor and protection)—albeit perhaps not quite as extensive as he claimed, and certainly not so that he could rule over it as a new country. But, still, while pointing at a slice of land and saying "That's mine now" on a flimsy legal basis, then persuading some impressionable folks to go and settle there at the risk of their very likely death certainly counts as fraud on a grand scale, it's also not *radically* different from how rather a lot of colonialism worked.

If MacGregor's harebrained scheme managed to convince a surprising number of people, it's because it was as much a reflection of the time as an aberration from it.

That doesn't fully answer the question, though—because it's not like there wasn't plenty of skepticism about MacGregor's claims to go round. There was, of course, Rafter's revenge-biography of him, written a year before the Poyais scheme got off the ground. This clearly tipped some people off to the fact that MacGregor was a wrong 'un. But the Poyais venture itself, and especially the book published to support it, received plenty of raised eyebrows in the press.

The London *Literary Gazette* of February 1, 1823, reviewed *Sketch of the Mosquito Shore* and had a few questions about some "remarkable peculiarities" in the descriptions of the country, such as "how rivers run *up* or how their course can extend hundreds of miles beyond the extreme breadth of the country."

"The whole thing," the *Gazette* sniffed, "smacks strongly of the Piratical and Buccaneering affairs of two centuries old."[117]

Particularly brutal was the *Quarterly Review*, a publication known for mixing Tory politics with literary reviews of almost unprecedented savagery. Their write-up of *Sketch of the Mosquito Shore* in the October 1822 issue is, shall we say, not kind. But to put it in perspective, just one year earlier Percy Bysshe Shelley had accused them of giving John Keats such a harsh review that it *literally killed him*.[118] Even taking into account the heightened sensitivity of Romantic poets, that's still quite an achievement in the field of nasty criticism, and suggests that, if anything, MacGregor got off lightly.

What's most notable about the *Quarterly Review* response is not merely that they're skeptical. For sure, they denounce the scheme's principals as "loan-jobbers and land-jobbers," joke sarcastically about a land "where all manner of grain grows without sowing, and the most delicious fruits without planting, where cows and horses support themselves, and where…roasted pigs run about with forks in their backs, crying, 'come eat me!'" before finally suggesting that "the whole affair [may be] merely, what is vulgarly called, a hoax."[119]

No, the really interesting thing is that the anonymous reviewer in question knows—in extremely fine detail—precisely what they're talking about. In modern parlance, they've got *receipts*. And also map references. "We must inform them…that Poyais is a paltry 'town' of huts and log-houses, belonging to Spain," they note, accurately, before going into several pages of minutiae about the local political situation and the exact nature of the treaties governing the region, all of which invalidated any claim MacGregor might have had to the land. They predict that "the settlers, if any such egregious simpletons should be found, will be considered…as trespassers, and treated accordingly." They question whether the book's author, Captain Strangeways, even exists, and say that, even if he does, nothing in the book gives any evidence of "his having ever set foot on" the Mosquito Shore. Finally, they wonder if Cazique MacGregor might be the same man who, a few years earlier, "being taken by surprise, jumped out of a window, his purse in hand, leaving his breeches behind him."

By any standards, you'd think that this should have thoroughly cooked MacGregor's goose; it's the kind of critical notice that's hard to come back from. And yet the Poyais scheme seems to have been virtually critic-proof.

We must confront two possibilities: firstly, that many of the settlers might not have been enthusiastic subscribers to the *Quarterly Review*; and, secondly, that they all fundamentally just *really, really* wanted MacGregor's fictions to be true. Which is a pretty powerful force, as extremely public grifters still know to this day.

This desperate desire to believe the fantasy also had staying power. Remarkably, a small group of the duped settlers continued to maintain for years afterward that MacGregor had been blameless in the whole affair, that everything was actually Colonel Hall's fault and that MacGregor had never actually mentioned many of the wilder things he was accused of promising the colonists—that was all down to the overactive imaginations of people like Picken. (This defense doesn't really stack up; while the settlers may certainly have talked themselves into believing more than even MacGregor sold them, it's hard to overlook the explicitly fraudulent details that MacGregor put into print—like how investors could claim their land by "presenting the title deeds at the Register Office in the town of St Joseph's, Poyais.")

But it wasn't just the victims of MacGregor's scheme who desperately wanted it to be true. Despite all the overt elements of scam, the central question of MacGregor's story remains: How much was he a con man, and how much did he truly believe? This question is especially persistent when you consider what happened after his scheme was exposed and he became a punch line to every joke in town, which is this: he simply carried on as if very little had happened.

He doesn't seem to have shown any remorse for the deaths of the people who trusted him—his only acknowledgment of the stories the survivors told when they got home was to sue the *Morning Herald* newspaper for libel after they printed a report of the survivors' testimony. He lost without ever turning up to court, because he'd fled to France—where he immediately set about trying to sell the Poyais scheme all over again.

In 1825, the London stock market crashed as the Latin American bubble burst, triggered in large part by the Poyais affair. Over sixty banks went under, the Bank of England had to be bailed out by the French and the effects were felt around the globe. MacGregor, meanwhile, was in France, writing the Constitution of Poyais and recruiting a new band of settlers. The French authorities caught wind of what he was doing only when they received an unusual number of passport requests from people wanting to travel to a country that didn't appear on any maps. MacGregor was arrested and charged with fraud, but the trial collapsed.

In total, MacGregor would subsequently spend over a decade more of his life trying to get his Poyais scheme off the ground, continuing to pursue it long after there was any chance of it being a successful con.

Tamar Frankel, a professor at Boston University School of Law, studied the profile of financial con artists in her 2012 book *The Ponzi Scheme Puzzle*. Some of the traits she identifies are unsurprising: con artists are lacking in empathy, narcissistic, greedy and self-justifying. When caught, they will deny and deflect, blaming just about anybody else rather than taking responsibility. They often justify their actions with the belief that they're simply reflecting the behavior of others: everyone else is crooked, too, and the victims deserved it because they were equally greedy and corrupt. "You can't con an honest man," as the saying goes. (Which is nonsense. You absolutely can! Some honest people are total suckers.)

But that's not all. In addition, con artists often have what Frankel calls an "addiction to unrealistic dreams and overwhelming ambitions";[120] comparing the skills of the con artist to that of an actor, she suggests, "It may well be that con artists act the character they have long been dreaming of."[121] MacGregor's dreams of an entire country may have been a little bit more unrealistic and overwhelming than many others...but the fundamentals were the same.

That apparently genuine belief in their schemes doesn't just help explain the actions of con artists themselves; it's also part of the reason why people trust them. "Their belief," Frankel writes, "can make them believable."[122]

Today, thanks to decades of movies and TV shows that delight in depicting the perfect con, we have the idea that all scams have to operate at a mind-bending level of complexity, filled with twists and turns and unexpected double-crosses. So it's perhaps worth remembering where we get the word from. The term "con" may be a generic one now, but its origins are very specific. The phrase "confidence man" was initially used about just one person: a chap named William Thompson.

Thompson was a scammer who operated in New York in the late 1840s, and his scam was one of glorious simplicity. A well-dressed and suave gentleman, Thompson would approach strangers in the street, strike up a genial conversation with them and then ask them the following question: "Have you confidence in me to trust me with your watch until tomorrow?"[123]

Thrown by the unexpected request, plenty of people would simply comply. Whereupon Thompson would...take off with their watch. Outstanding.

Thompson may have been the first to get the title of "con man," but of course there have been swindlers ever since there have been suckers, which is to say forever. Perhaps America's first truly legendary grifter was Tom Bell, who operated in the first half of the eighteenth century. Having been expelled from Harvard following a history of "saucy behavior," he used his knowledge of the social cues of America's wealthy elite to scam his way across the colony for many years, ruthlessly exploiting the assumption that someone with nice clothes and the airs of the upper classes couldn't possibly be a crook. (He may well also have been the scammer who, in the guise of a schoolmaster named William Lloyd, stole a ruffled shirt and a handkerchief from Benjamin Franklin after talking his way into Franklin's house. That's not an especially inventive con—I'd just started to feel uncomfortable because we hadn't mentioned Franklin in a while.)

If you'd like an example of a genuinely complicated scam with some far-reaching consequences, you can't do much better than Jeanne de Valois-Saint-Rémy, "Comtesse de la Motte," a social-climbing French con artist who parlayed her self-bestowed title and entirely invented friendship with Marie Antoinette into a scam to buy a priceless diamond necklace with borrowed money—a scheme that at one point involved her hiring a prostitute to impersonate the queen in a meeting with a Catholic cardinal (who Jeanne was also having an affair with). The scam nearly came off but failed when the queen got word of it. While Jeanne was tried, convicted and jailed for the crime, it didn't work out terribly well for Marie Antoinette either; the trial focused public attention on the lavish spending of the royal family, turning her from a not terribly popular queen into a wildly unpopular one. All of which would help spark the French Revolution a few years later and led to Marie's eventual appointment with the guillotine.

But, while these grifters were motivated largely by wealth,

Jeanne de Valois-Saint-Rémy, or as she liked to call herself, Comtesse de la Motte.

one of the most fascinating con artists of all time was motivated by something very different.

This tale begins in the autumn of 1951, in Edmundston, Canada, when Mary Cyr picked up a newspaper and was surprised to read a story about how her son, Joseph, was a war hero.

The paper recounted how, serving with the Canadian navy in the Korean War, Dr. Joseph C. Cyr had saved the lives of many badly wounded South Korean soldiers. They'd been picked up in a small boat, on the verge of death—but, performing emergency surgery right through the night in an improvised operating theater while the ship rode out a fierce storm, including removing a bullet from right next to one man's heart, Dr. Cyr had brought them all through safely. Happy to have some good news from the war, the Canadian military's press office was proudly trumpeting his selfless heroism and skill.

The reason this all came as a surprise to Mary was because she was pretty sure her son wasn't in the Canadian navy, and he almost certainly wasn't somewhere off the coast of Korea. In fact, he was supposed to be in general practice, about forty miles down the road. Still, she thought it best to double-check.

It was perhaps unsurprising that Joseph Cyr—an easygoing and kind man, who was fluently bilingual, having an anglophone mother and a francophone father[124]—had wound up practicing medicine in the small, extremely bilingually named New Brunswick community of Grand Falls / Grand-Sault, located a stone's throw from Canada's border with Maine. And that's exactly where he was, quietly minding his own business, when he started getting the phone calls asking him if he was on a ship in Korea.

His initial instinct—to pass it off as a simple case of mistaken identity, someone with the same name—quickly foundered when it turned out that he was the only Dr. Joseph C. Cyr in Canada. He remembered that several of his medical diplomas and other identifying documents had gone missing the previous winter. And thinking about it, he knew who had taken them—Brother John, a local monk he had become good friends with, shortly before Brother John mysteriously vanished.

Brother John, of course, was not actually Brother John. Nor was he the biologist and cancer researcher Dr. Cecil B. Hamann, the identity Brother John had gone by before turning to the Church. He also wasn't Dr. Robert Linton French, the Stanford-educated psychologist he had previously been, before becoming Drs. Hamann and Cyr.

Brother John was in fact an American named Ferdinand Waldo Demara, a man who would soon become immortalized as "the Great Impostor." Demara stands out among the ranks of legendary con artists by virtue of apparently not being particularly motivated by financial gain. I mean, sure, he passed his fair share of bad checks and abused the odd expense account during his career, but he never used his undoubted skills to pursue vast quantities of wealth or a jet-set lifestyle. He had a talent for social engineering, persuading bureaucracies to give up people's identity documents and convincing people from all walks of life to put him in positions of trust—a talent that he could have used for terrible ends, but which he actually used almost exclusively to inveigle himself into a series of perfectly worthy and upright public-service jobs. Over the years, he would be a doctor, a sheriff's deputy, a law student, a prison warden, a lot of teachers and a wide variety of monks, and he would found both a college philosophy department and a whole university.

Demara didn't con people to get their money. He conned people to get their respect—and, perhaps, his own too.

The thing about Demara is that he wasn't just good at scamming his way into jobs; he was often surprisingly good at doing them too. He was a lightning-fast learner with a near-as-dammit photographic memory. As Dr. French, he managed to convince a Catholic college in Pennsylvania to make him the dean of their new school of philosophy and would go on to teach psychology at another Catholic college (his secret, he said, was doing the reading just ahead of the class—"the best way to learn anything is to teach it"[125]). As Cecil B. Hamann / Brother John, lacking a college to teach at while studying as a novice with the Brothers of Christian Instruction, he managed to railroad both the monks and the local authorities into founding a private university—only to storm off in a fit of pique when they gave it a name he didn't like. (The university he founded still exists today, after a change of both name and location, as Walsh University in Ohio.) And, of course, there was his miraculous, lifesaving work on board HMCS *Cayuga* as Dr. Cyr, which he managed by sneaking into his cabin and speed-reading a surgical textbook shortly before operating.

Those kind of prodigious talents would have served him perfectly well under his own name—would have made him famous, even—but Demara seems to have never felt comfortable as himself. He was trying to find his place in the world, and becoming somebody else—particularly somebody with credentials that Demara lacked—seemed to offer a shortcut past the boredom and frustrations of having to navigate life in the slow lane.

He found it hard to settle, never entirely sure which of his many personas he truly wanted to be. He would return to teaching repeatedly; he tried enlisting in (and going AWOL from) the military several times under various guises; his numerous efforts to take religious orders, both as himself and under a suite of aliases, seem to have been at least partly based on a genuine desire for spiritual development. His search for a sense of vocation reads like a fun house–mirror version of people in their twenties bouncing between careers as they try to figure out who they are. (Note for younger readers: this was a thing you could do before the crash of 2008. It was nifty!)

After he was exposed for not being Joseph Cyr in 1951, his case became a sensation across North America. In 1952, he granted a lengthy interview to *Life* magazine in which he told his (quite possibly unreliable) version of the story, at the end of which he expressed his desire to maybe, finally, just be himself.

He would express the same desire to the press in 1956,[126] when he was arrested following his stint as Benjamin W. Jones, a prison warden in Texas—a job that came to an end when one of the prisoners recognized him from an old copy of *Life* magazine. That plan to settle down as plain old Ferdinand lasted barely a few months before, suddenly, he was Martin Godgart, a teacher at a needy school on a remote island in Maine. After he was arrested there, he told his story again, this time to the author Robert Crichton, insisting that he was definitely going to go straight. Sometime after that, he was Godgart again, this time teaching indigenous children in Point Barrow, Alaska, at the very tip of the northernmost point of the United States— the most remote place imaginable, as though he was trying to physically run as far as possible from his past. That was all going terribly well until a passing trapper recognized him from *Life* magazine again. After that, he tried being a bridge engineer in Mexico and a prison governor in Cuba, with limited success.

Crichton would go on to turn his story into the bestselling book *The Great Impostor*, which itself would become a light comedy film starring Tony Curtis. Demara was unhappy. He complained that the film took liberties with the truth.

By this point, Demara had reached such a level of fame that he could no longer pass himself off as anybody else. From 1960 onward, trapped by his own notoriety, he was forced to live in the prison of himself. He finally took religious orders once more and became a pastor—under his own name, this time—living out another two decades of good, generous life in a loving community, as Ferdinand Waldo Demara. When he died, in 1982, his doctor told the AP, "He was about the most miserable, unhappy man I have known."[127]

Demara was able to flit between identities and establish himself in positions of responsibility so easily because he exploited

structural qualities of American society at the time. His progress was eased by a flurry of letters of recommendation from
sundry bishops and other notables (all written for the person he
was impersonating), all of which were taken on trust and seen as
validating his identity. Once he had a foot in the door, he knew
what levers to pull to cement his position. As Crichton put it in
The Great Impostor, Demara's key insight was that "in any organization there is always a lot of loose, unused power lying about
which can be picked up without alienating anyone."[128] Which,
honestly, would work just as well as the basis for a corporate
self-help book on how to get ahead in the business world as it
does for the biography of a con man.

Good con artists adapt to—indeed, are products of—the cultures in which they operate. Where Demara found the loopholes
in 1950s USA, Vladimir Gromov did the same for the Soviet
state of the 1920s and '30s.

Stalin's Soviet Union might not immediately seem like an
ideal place to earn a living as a con artist, and, indeed, if you
judge Gromov's life by petty metrics like whether he was sen-
tenced to death at the age of thirty-six, then he might have been
wiser to seek an alternative profession. On the other hand, if you
judge it by whether he managed to get his death sentence commuted by writing a play about a love affair between a Bolshevik man and a beautiful capitalist woman half his age, which he
sent to the deputy procurator-general of the USSR, then things
look a bit rosier for Gromov.

Gromov's insight was that the climate of fear, oppressive bureaucracy and ideological rigidity of the early Stalin years was
in fact ripe for exploitation—which he managed to do in spades,
appearing in a wide variety of guises as an expert engineer or
an acclaimed architect, amassing a small fortune along the way.

He realized that the Soviet bureaucracy's unquenchable hunger for documentation left the system with very little capacity to
actually check the validity of the reams of paper they were ac-

cumulating. So instead of trying to fly under the radar, he opted to flood the system, stealing or forging documents with wild abandon to enable him to hop between "jobs." Any questions he might face could be deflected with the proper appeals to Bolshevik dogma, and, once he had persuaded somebody to accept his identity, he exploited the intimidating power of status under Stalin to ensure that nobody else would question him—a perfect bullshit feedback loop, enforced by the very authoritarian culture that was supposed to stamp out transgression. In the words of the historian Golfo Alexopoulos, "He did not avoid the authorities but bombarded them with false employment papers, phony requests for money and goods, and vicious denunciations."[129]

His standard modus operandi was to establish phony credentials with the help of fake documents and to use that to get himself an appointment to a senior role in a state industry—ideally one in a far-flung location somewhere in the Soviet empire. He would obviously need his wages advanced and his travel expenses paid up front. By the time the coal mine in Vladivostok realized that their chief engineer had never turned up, Gromov would be somewhere else, already starting on a new "job."

The crowning achievement of Gromov's scamming came when he managed to get himself appointed to the exalted role of engineer-architect for a major new fish cannery near the Kazakhstan-China border. Now, this might not immediately strike you as the world's most glamorous assignment, but in the Soviet Union in the 1930s, it was a really big deal. So much so that Gromov managed to persuade the commissar of supply, Anastas Mikoian, to send him the vast sum of one million rubles, through the cunning technique of asking for two million rubles. (To give you a sense of how much that was, the average yearly salary at the time was just over 1,500 rubles.[130])

The Kazakh cannery was the peak of his career, but, unfortunately for Gromov, it was also where his downfall occurred. That's because he made the classic mistake of abandoning his

tried-and-trusted methods—namely, the method of bolting before anybody sniffed him out. This time, Gromov felt that he was
onto such a good thing that he decided to stay on the ride and
fully embrace his false identity as an engineer-architect. Possibly,
much like Demara, he just wanted to put down some roots and
actually become the person he was pretending to be. Maybe the
power and money went to his head. Alexopoulos suggests that
"perhaps Gromov ceased to be an imposter by 1934, not because
he had internalized or come to believe his own lies, but because
he viewed those around him at Glavryba's monumental construction project as essentially no different from himself."[131] In
other words, if everybody else was faking it, why shouldn't he?

But this desire to settle down into his newfound status couldn't
actually survive too much contact with the harsh reality that,
bluntly, he didn't actually have many skills in the fields of engineering, architecture or fish canning. Gromov's tactics of de-
nouncing anybody who questioned him as an enemy of Stalinism
were effective in the short-term, when he was constantly on the
move…but, after too long in one place, they merely built up a
critical mass of people with an enormous grudge against him.

But even after his arrest and death sentence, he still managed
to escape, one last time, turning the creative energies that had
fueled his production of imaginary work orders, invoices and
telegrams into a more traditional form of fiction.

His prison-produced play, titled *Love and Motherland*, may not
have been terribly good. In fact, when the procurators passed
it on to the head of the playwrights' union for a professional
opinion of Gromov's literary merits, it received the kind of critical notice that would terrify any author, never mind one who
was relying on a single manuscript to save him from execution.
The union man wrote that Gromov's "playwriting ability is extremely low" and that "the work has no ideological or artistic
value and is clearly unacceptable from any standpoint."[132] I think

it's fairly safe to say that John Keats would not have coped well with a review like that.

And yet, miraculously, it worked. Gromov's death sentence was commuted to ten years of labor. To this day, it remains unclear exactly what could possibly have convinced a senior Soviet official to spare Gromov's life simply on the basis of a play depicting a senior Soviet official as a handsome and heroic figure who bones a hot twenty-three-year-old Parisian woman and converts her to socialism through the power of his ideological and sexual magnetism. Ah, well—I guess that will just have to remain as one of those inscrutable historical mysteries we will never solve.

If this ability to find the loopholes in society and ruthlessly exploit them is the mark of a great con artist, then the star of our final story must stand as one of the greatest.

While Gregor MacGregor may have been the grifter in history who operated on the largest stage, this is the tale of a woman whose ambition and chutzpah matched Gregor's step for step, but she worked at the other end of the detail scale. Where Mac-Gregor's scam required him to invent a whole country, Thérèse Humbert's revolved entirely around the contents of one single locked safe—a prop that, through a gloriously simple piece of legal judo, enabled this formerly penniless country girl with an overactive imagination to spend two entire decades living a life of luxury in the Paris of the belle époque.

That safe contained, it was said, a set of bonds worth an estimated 100 million francs. These had supposedly been bequeathed to Thérèse by a mysterious American gentleman named Robert Henry Crawford, whose life she had saved on a train, some years before, when he had suffered a heart attack. Out of gratitude, he pledged to reward her handsomely—a promise he kept in his will, which he changed shortly before his death so that she would inherit much of his vast fortune.

On the basis of this imminent wealth, Thérèse was able to borrow money freely, as lenders gleefully anticipated the huge returns they would shortly get. It wasn't a complex scam; fundamentally, it's just the old "check's in the post" routine. And, of course, that trick works only for a limited time—because eventually the postman turns up empty-handed.

Thérèse Humbert knew this perfectly well, because she had a long history of inventing wealthy benefactors before eventually getting caught. Ever since she was a child, the line between her real life and her fantasy life had been blurry, verging on nonexistent. In this, she was influenced by her father, Auguste Daurignac, an eccentric and somewhat pathetic dreamer who believed himself descended from nobility. He spent his later years insisting that he could do magic and spiraling into debt while reassuring his creditors that he had documents proving he was due a vast inheritance, which were locked in an old chest.

Forced by her father's cognitive absence to run the household, Thérèse Daurignac took his fantasies and turned them into a practical, if temporary, source of income. Charming and utterly ingenuous, she ran up debts with virtually every tradesperson in the greater Toulouse area, all the while promising that they would be paid after a nonexistent inheritance came through, or after a fictional wedding to the scion of a great shipping family. As her biographer Hilary Spurling writes, "All her life Thérèse treated money as an illusion: a confidence or conjuring trick that had to be mastered."[133]

But as was bound to happen, that particular con ran out of road, and the Daurignacs were evicted from their home in a whirlwind of debt. But it wasn't long before they bounced back, powered once again by Thérèse's wild imagination. This time, it was one of her longest-standing fantasies that aided her cause: the Château de Marcotte, a grand mansion, far away on the coast, which she had long daydreamed about inhabiting.

Thérèse Humbert

The Château de Marcotte didn't actually exist, but that had never stopped Thérèse talking about it like it was the most real thing in the world. "She lied as the bird sings," as one acquaintance would later recall.[134] And so convinced—and convincing—was she in her descriptions of this luxurious property, with its marble floors and lush gardens, that she seems to have persuaded plenty of other people that it was real too. Included in that was her future father-in-law, Gustave Humbert, a senator and rising star of French politics. Humbert didn't just approve of his son Frédéric marrying Thérèse; he also gave the thumbs-up to his daughter, Alice, simultaneously marrying Thérèse's brother,

Emile, in a double wedding. Exactly why an ambitious (but not wealthy) politician would want to tie his family so closely to a bunch of weird, impoverished ne'er-do-wells like the Daurignacs is a bit of a mystery, until you factor in the possible lure of them having a very big house in the country.

Oh, also, the newlyweds were cousins. Humbert's wife was Thérèse's aunt. Little extra detail for you there.

With her new political connections, Thérèse was right back in the game, and, with the senator's help, quickly set about borrowing money against both her fictional château, as well as a fictional cork plantation in Portugal. But soon she wanted more, and so, in 1883, Robert Henry Crawford and his 100-million-franc will came onto the scene. The promise of that inheritance, and the money she could borrow in advance of it, would in itself have probably kept the Humberts going for a few years. But this is where Thérèse (quite possibly in collaboration with both her husband and her father-in-law) played her master stroke.

If the British weakness that MacGregor exploited was its predilection for colonial fantasy, and if the American weakness that Demara exploited was its reverence for credentials and its careless assignment of individual power, and if the Soviet weakness that Gromov exploited was its oppressive ideology and bureaucracy, then the French weakness that Humbert exploited was this: its shit-awful legal system. The French courts of the time were notorious for the slow, grinding way they proceeded and for their only fleeting attachment to notions of justice. It was in this context that Thérèse came up with a plan to extend the lifespan of her con—a plan so simple and so cunning that, frankly, I'm in awe.

She sued herself.

Or, more precisely, she invented a couple of fictional American nephews of her fictional American benefactor, in order that *they* could sue her, contesting the will. The point of this was not that they should win—in fact, the most important thing was that nobody ever be allowed to win, with every verdict leading to another appeal, and counter-appeal, and then around again, all at the slowest possible pace that the French courts could achieve. The Crawford brothers never even needed to make an appearance, instead instructing some of Paris's finest lawyers by letter from across the ocean. The only thing that mattered was that the case drag out indefinitely, so another year could go by in which Thérèse was kept from the inheritance she was permanently on the brink of acquiring and so would be forced to borrow vast sums of money from a swarm of eager lenders.

And, all that time, by the strict order of the court, the actual documents must be locked securely away in Thérèse's safe, never to be seen.

The Humberts managed to keep this going for twenty magnificent years: two decades in which they lived the most extravagant lives possible in Paris, at a time when the bar for extravagant Parisian living was really very high indeed. Thérèse and her husband occupied one of the most luxurious apartments on the avenue de la Grande-Armée; their parties were legendary and attended by all the great and the good of the time, from the actress Sarah Bernhardt to the president of the republic. Thérèse, a wide-eyed country girl from an impoverished family, was now one of the most influential women in France.

And, if occasionally one of Thérèse's creditors got anxious about the vast sums of money they had given her with no return, and started making threatening noises, well...most of the influential people they could have turned to for help enjoyed going to the Thérèse's parties.

The scheme's downfall, when it came, was abrupt, and stemmed from a simple and uncharacteristic slip: Thérèse made the mistake, when asked to provide an address for the Crawford brothers in New York, of simply making one up. She may have thought it would be too much effort for anybody to check and find that nobody called Crawford lived at 1302 Broadway, but when there were millions of francs involved and an increasingly upset army of creditors gathering, the effort barrier suddenly became much lower. A court finally grew suspicious and ordered that the will be examined.

And so it was that, on May 9, 1902, a huge crowd of up to ten thousand gathered on the avenue de la Grande-Armée to witness the famous safe being lowered from the Humberts' apartment and opened. After some effort, the appointed locksmiths finally forced it open with hammers. The crowd peered eagerly inside, hoping to finally get a glimpse of the vast riches contained within.

They were rather surprised when the contents of the safe were revealed to be, in total, "an old newspaper, an Italian penny, and a trouser button."[135]

The famous safe being lowered from Thérèse's home.

Thérèse Humbert managed to convert that newspaper, coin and button into decades of luxury simply because she had an instinctive knack for how to exploit the weaknesses in her fellow humans and the social systems they'd created.

As a friend of hers, who wrote anonymously under the very cool name of "Madame X," put it:

> *What demonstrates above all else the genius of Thérèse is the grandeur, the sheer immensity of the scale on which she operated. If she had laid claim to an inheritance of no more than four or six million, she would not have lasted two years, and would with difficulty have managed to raise a miserable few thousand francs. But a hundred million! People took their hats off to a sum like that as they would have done before the Pyramid of Cheops, and their admiration prevented them from seeing straight.*[136]

6

LYING IN STATE

If there's one thing everybody knows about politicians, it's that they lie. They lie about big things, and they lie about small things, and they lie about all the sizes of things in between. Surveys of the most and least trusted professions regularly put politicians right at the bottom, below real estate agents and even—God forbid—journalists. As the well-worn joke has it, you can tell when a politician is lying because their lips are moving.

Here's the thing, though: most politicians actually don't lie anywhere near as much as you think! I know this may sound somewhat implausible, especially given [waves vaguely at the world in general] *recent events*, but trust me—fact-checking what politicians say is very literally my job. Lies are actually a far, far smaller part of the day-to-day business of politics than the stereotype would have you believe.

This isn't to say that politicians (and our leaders in general, and indeed the whole apparatus of the state) are universally noble, upstanding and entirely trustworthy individuals, selflessly committed to truth telling under all circumstances. That's...well, it's obviously absurd, although perhaps not *much* more misleading than the notion that politics is naught but a writhing snake pit of eternal deceit. But the point stands: if you think that politics is no more than the business of telling the most convincing lies, then you've got a skewed view of how we're governed.

Our leaders do lie, of course. Just as with the human population at large, a small number of them lie habitually—they lie as a first recourse rather than as a final desperate fallback, and they often seem to actively enjoy lying. You can probably come up with more than a few examples of politicians in this category off the top of your head right now (which ones you choose will probably depend on your own political preferences).

But most of them lie only occasionally, if at all, and when they do it's very often for the same sorts of dumb, basic reasons the rest of us lie: to get out of an awkward conversation, to disguise the fact that we are basically winging it in our job or to hide that we're having sex with someone who—for whatever reason—we don't want to publicly acknowledge we're having sex with.

There's a reason that "it's not the crime, it's the cover-up" is a cliché, which is that, quite a lot of the time, what brings politicians down are the lies they tell to stop people finding out about things that would be, at most, a bit embarrassing. (That said, the phrase is thought to have originated with the Watergate scandal—where, as we'll discuss briefly in a bit, while there was certainly a cover-up, there was also a shitload of crimes.)

So why are lying and politics so inextricably linked in our minds? The problem is twofold. The first issue is that, if the business of politics doesn't necessarily attract a higher proportion of compulsive liars than other careers (I am aware of no research on this—someone please do it), it certainly offers those who do have that tendency more than ample opportunity to practice their craft in an extremely public way. In a way that might not be the case if, say, they worked in a small agricultural transport firm in Iowa.

A politician is daily offered the opportunity to lie about six things before breakfast—and, more important, they'll probably find both a willing platform and a receptive audience for those lies. There's always someone out there who wants to hear comforting or infuriating deceit: that we're entering a glorious new age, or that there's someone else to blame for our problems, or that the world isn't complicated and gray, but simple and monochrome. (And if you think that previous sentence was talking about *those other people* and not you, it was probably talking about you.)

And secondly, if you lie when you're in charge of a country, it *really fucking matters*.

For starters, it means that, in those marginal decisions we all have to make between taking the honest path and the dishonest one, politicians often have a far greater disincentive to be honest. If you forgot to reply to an email in your job at Big Jim's Pig Wranglers in Iowa, then some pigs might get stuck near Crawfordsville. That's bad news for the pig farmer and might lose your firm a bit of business; you'll almost certainly have to do some kind of "sorry I let the team down" all-staff email. But the incentives are still probably on the side of fessing up and taking the flak. By contrast, if you forgot to reply to an email in your job overseeing the TSA, then several million extremely angry voters might get stuck at airports shortly before Thanksgiving, cable news will be gearing up for a full-on ragefest, and an email saying "Hey, to err is human; I hope we can move beyond this" probably isn't quite going to cut it. We all say that we'd like politicians to be more honest, but generally we haven't shown much indication that we're prepared to repay them when that honesty involves saying, "Holy wow—yeah, I screwed the pooch on that one. I've learned a lot from this and I'll do better next time."

Also, when our leaders lie, sometimes really, really, *really* large numbers of people die. There can be wars and stuff. And, yeah, that sort of thing does tend to stick in the mind a little.

Political lying has been with us for as long as, well, politics (exactly when we invented politics is unclear, but it's safe to say it was quite a long time ago). To take just one example, one of history's most notable liars was a chap named Titus Oates, who, in 1678, managed to send England and Scotland into a state of wild anti-Catholic hysteria for three years on the basis of some extremely transparent lies.

Now, it's important to not overstate how unusual that was. For much of our history, getting the British to indulge in anti-Catholic hysteria has been roughly as difficult as getting a dog to freak out about its own tail. But, still, it's notable that, for several years, the country's most influential people were in thrall to a man who'd got himself ordained as a vicar by falsely claiming to have a Cambridge degree and then had spent most of the next decade fleeing various charges of perjury and buggery.

Once described as "the most illiterate dunce, incapable of improvement,"[137] Oates was a dull, unhappy child, living in the shadow of a violent father, and was once expelled from school for fudging his tuition fees. He tried to study at two different Cambridge colleges, but never graduated—although, while at Cambridge, he did gain, in the words of the *Oxford Dictionary of National Biography*, "a further reputation for stupidity, homosexuality, and a 'Canting Fanatical way.'"[138] (I mean, not that any of those things are particularly unprecedented at Cambridge.)

In 1677, after a brief stint as a Royal Navy chaplain that ended rapidly when he was accused of doing gay stuff, and at least two escapes from prison over a perjury charge, Oates decided that this would be a good time to convert to Catholicism. Helpfully, at the same time, he also fell in with a quite probably insane anti-Catholic conspiracy theorist named Israel Tonge. That, er, unusual combination of influences would set Oates up perfectly for his most infamous contribution to history: the spurious allegation that there was a Popish Plot to assassinate King Charles II.

Titus Oates: incapable of improvement.

This involved writing a dense, sixty-eight-page pamphlet filled with wild allegations of plots and the names of over a hundred conspirators, which Oates and Tonge planted in the house of Sir Richard Baker—a fellow anti-Catholic zealot—where Tonge promptly and fortuitously "discovered" it the next day. No, it doesn't make sense. Why would it be there? Why would the Catholics write their plot down and then accidentally leave it in the house of someone who hated them? Look, conspiracy theories have never had to be logically consistent, right?

Tonge then had a friend approach the king to warn him of the plot. It's worth noting that Charles II was absolutely having none of it; he thought it was all nonsense. But the same can't be said of Charles's ministers, or Parliament in general, who absolutely ate it up. Oates was called to testify before the Privy Council, where—despite Charles II himself cross-examining him skeptically—the politicos decided he was telling the truth. Whenever Oates hit a snag in his narrative, his solution was simply to invent whole new plots and accuse more people; he told the assembled dignitaries exactly what they wanted to hear, and the fact that his claims weren't even internally consistent didn't seem to matter much. At one point, the king caught him in a particularly brazen lie and had him arrested; Parliament overruled this decision and not only released Oates but gave him a house and a salary. Among the people that Oates would end up accusing of plotting to murder the king were the queen (a Portuguese Catholic, who wasn't exactly popular in England), the diarist Samuel Pepys and the schoolmaster who had expelled him years earlier.

The result was complete hysteria. Scores of prominent Catholics were arrested and put on trial—twenty-two were executed. Catholics were expelled from London. The media and the public stoked the fears and contributed their own inventions, with panics about Catholic plots and suspicious figures spreading like wildfire. It was several years before the hysteria would die down and Oates would come to be viewed with skepticism, eventually being asked to move out of his government-provided house, as everybody got a bit embarrassed by the whole thing.

How did Oates, a person with a terrible reputation and an incoherent story, acting alongside a probable lunatic, come to control the political narrative of an entire country for several years—when even the person who was supposedly the target of the assassination plot didn't believe him? Like many modern conspiracy theories, it played to an agenda that lots of people already had: they wanted to believe, and that meant its contradictions and inconsistencies didn't hurt it. But there was also Oates himself, an unattractive and uncouth man, who nonetheless seemed to have a magnetic sway over his listeners. To put it simply, he was a deeply talented bullshitter, who, even when he wasn't plausible, was at least entertaining. As the writer Sir John Pollock put it, "His gross personality had in it a comic strain. He could not only invent but, when unexpected events occurred, adapt them on the instant to his own end. His coarse tongue was not without a kind of wit. Whenever he appears on the scene...we may be sure of good sport."[139]

You can't talk about political lying without mentioning Watergate, but I kind of feel that, with multiple Hollywood movies on the subject, this has been covered enough. I'm guessing you roughly know the story? I mean, if you don't, look it up—it's a doozy. But, still, there are a couple of aspects that are worth revisiting. Probably the second most interesting thing about Watergate is how very, very close they all came to getting away with it. The *Washington Post* articles that played the major role in uncovering it all were a slow drip-drip of mostly non-earthshaking stories, and it could very easily have gone the other way—where people just took the previous revelations for granted, adjusted their internal how-much-dishonesty-is-too-much-dishonesty bar, and so the story never quite swelled into the world-shaking scandal it became.

The most interesting thing is how astonishingly bad they all were at lying.

I mean, really awful—just gobsmackingly incompetent. For starters, you have the basic, famous fact that Nixon recorded all of the conversations in the Oval Office, where they talked about the bad things they were doing. Nixon wasn't the first president to bug his own conversations—that was FDR—but he was the first to do it as a matter of routine, which is weird when you consider that he was probably discussing way worse stuff than most other occupants of that office. (I mean…maybe he wasn't. How would we know?) The fact that the most plausible explanation offered for this behavior to date comes from *Doctor Who*—namely, that he was doing it to combat the effects of memory-erasing aliens—maybe gives a sense of how supremely dumb and inexplicable this was.

But the really good stuff is about the eighteen and a half minutes of tape that we don't have. That's the total amount of time that was "accidentally" erased from the tapes covering the morning of June 20, 1972—during a conversation between Nixon and his chief of staff, H. R. Haldeman, three days after the bungled Watergate office break-in. Given that the tapes that *weren't* erased were still enough to condemn Nixon, one can only assume that the erased portions must have contained something approximating the following exchange:

NIXON: So, can you update me on the crimes we are doing?

HALDEMAN: Yes, uh… Yes, the crimes.

NIXON: The crimes—how are they? The crimes I ordered you to do. Tell me about the crimes.

HALDEMAN: The crimes, uh [inaudible], crimes have happened. We did the crimes, as you very specifically demanded.

NIXON: Okay, good. I am glad about the crimes—the crimes I explicitly told you to do and that you then agreed to do. It is good that the crimes have happened. [inaudible] Hot diggity, I love those crimes.

HALDEMAN: Yes, but the crimes went wrong. They found out about the crimes. This is bad.

NIXON: Oh, no. Now we must do more crimes to stop people finding out about the previous crimes.

HALDEMAN: Yes, okay... Okay, yes, more crimes. Understood. Together, let's do crimes immediately.

NIXON: Okay, yes, good. Thank you for the crimes—the crimes we share. [inaudible] I hate communists. Hoo, boy, do I crave alcohol.

The best bit of the failed Watergate cover-up comes from the wonderfully bad attempts to explain why the tapes had been wiped. Nixon's secretary, Rose Mary Woods, publicly took the blame for "accidentally" wiping the tape. She had been transcribing the tape, she said, when she was interrupted by a phone call. Reaching over to answer the phone, she accidentally pressed Record on the tape machine and kept her foot on the pedal that made the tape go forward, all the way through the five-minute call. Ignore for a moment the fact that this doesn't explain the other missing thirteen minutes, that the wiped portions weren't in a single block, but four or five different sections, and that the model of tape machine in question didn't work like that.[140] Focus instead on the fact that someone decided it would be a good idea to get Ms. Woods to demonstrate for press photographers exactly how she had come to accidentally erase the tapes, to illustrate just how credible and plausible her story was.

Behold, Rose Mary Woods's extremely normal phone-answering-while-keeping-foot-on-pedal-for-five-minutes pose:

Rose Mary Woods, reaching.

Dubbed by the press "the Rose Mary Stretch," there's a plausible case to be made that—for all the heroic investigative reporting of Woodward and Bernstein—it was the Twister-like image of a middle-aged woman extending herself as far as she could to try to reach the phone and the pedal at the same time that really made the American public think, "Hmm, something's not right, here."

If there's a time when the dishonesty of our leaders truly comes into its own, it's when somebody wants to go to war. A really large number of wars were sparked by inciting incidents that, in retrospect, turned out to have been less than accurately reported. There's the second Gulf of Tonkin incident that provided the justification for the Vietnam War, which subsequently turned out to have involved a completely fictitious attack on a US boat. The Spanish-American War of 1898 had a major catalyst in the sinking of USS *Maine* in Havana, which a rabid American press blamed on the Spanish—even though it was initially believed to be an accident, and subsequent investigations mostly agree that the likely cause was a spontaneous fire in a coal bunker. And, of course, there's that thing about Iraq having weapons of mass destruction that could be deployed in forty-five minutes.

At the top of the roll of dishonor that is bullshit reasons for trying to start a war, though, must come the Suez Crisis. At the time of writing, in the UK, Suez is having something of a rhetorical renaissance, thanks to its frequent deployment in phrases such as "This is the worst crisis since Suez." The current state of British politics doesn't actually bear any real resemblance to Suez (I mean, for one thing, it involved cooperating with the French), but it's still worth revisiting briefly, if only to note how the crisis played out: a nation was left humiliated, and a prime minister ended up resigning without people even knowing *quite* how much bullshit had been involved.

In short: it was 1956, the Age of Empire was ending and Britain was not coping well with the breakup. Rather than taking the time-honored post-dumping approach of sitting under a blanket, binge eating and listening to Alanis Morissette, the UK decided to do a war instead. Having belatedly withdrawn from Egypt, the Brits were rather miffed when Colonel Nasser seized power in a coup and promptly nationalized the Suez Canal— the vital trade route between the Red Sea and the Mediterranean that, until the nationalization, had been jointly owned by the UK and France.

The question was what to do about it. In Britain, Prime Minister Anthony Eden was being urged to take a hard line, especially by some in the press—*The Times*, in particular, possibly remembering its ill-fated support for appeasement prior to the Second World War, urged Eden to get tough. A similar story was playing out in France. But it wasn't clear that military action would (a) work, or (b) be terribly popular with any other countries. Nasser's actions might have been annoying, but it wasn't clear they were illegal: the shareholders of the canal company had been paid at market rates. So, for several months, the situation remained in a tense, nervous limbo.

This all changed at the end of October, when Israeli forces invaded Egypt. This obviously raised the possibility of a huge and bloody war that could envelop the whole Middle East; quickly, British and French troops moved to intervene as peacekeepers and separate the Israeli and Egyptian forces. Which, entirely coincidentally, meant that they would also have control of the canal.

Some people couldn't help but find all this a little bit...*convenient*.

In Britain, the mood began to shift. While the war still had support, the tough stance that previously had overwhelming approval from across the political spectrum began to attract more and more criticism. The mood of the press changed; *The Times* started to urge caution, while the *Manchester Guardian* came right out and suggested there was something dodgy going on. The international response was worse: condemnation from around the globe and, in a profound blow to Eden's plans, a blunt refusal from the USA to support the venture, and the threat of oil sanctions if they carried on. Yes, America opposing a war in the Middle East. It was a different time.

The result of this wild miscalculation about Britain's post-imperial ability to impose its will on the world was a humiliating withdrawal some weeks later. Eden insisted to Parliament that the UK had had no foreknowledge of Israel's invasion, but his authority was gone and his health was failing. He resigned in January 1957.

The thing is, all of this happened without people knowing the full story, despite the widespread suspicions. That wouldn't come out for several decades, when it was finally revealed that Britain hadn't just had foreknowledge of the invasion—they'd planned it. Eden's denials had been complete horseshit. In fact, Britain, France and Israel had secretly plotted every stage of the war in advance—Israel's invasion, and the "peacekeeping" response. This was all decided a week before, at a covert meeting in France, where the three parties had drawn up a document outlining exactly what role each of them would play in this geopolitical pantomime. Britain destroyed their copy of the document. Unfortunately for Eden's historical reputation, Israel kept hold of theirs, quite reasonably not trusting the two European countries to keep up their side of the bargain.

This also explained the somewhat unexpected shift of mood in the pages of *The Times*: their senior editors had been briefed by the government on the plans for the war before it all happened.[141] Realizing that this was a terrible idea, they quickly shifted their line. Of course, they didn't actually think of *reporting* the fact that they knew the war was based on a lie.

It's not just in the beginnings of war that the falsehoods of statecraft flourish. Rather notoriously, wars are not great for producing reliable information: the fog of war means that many of the details coming from the battlefield are unreliable at best. But, more than that, war provides a tinderbox for the blend of rumor, myth and propaganda that fuels wild and uncontrollable falsehoods.

You can see this in all those reports from the field during the First World War that so annoyed H. L. Mencken a few chapters ago. While his estimate that 99 percent of all war reporting at that time was bullshit is probably a tad overstated, that unprecedented conflict gave rise to a legion of completely untrue stories.

There was the widespread story of the Canadian officer who had been bloodily crucified by German troops near Ypres, pinned by bayonets through his arms and legs. The details varied: when it was reported in *The Times*, he had been pinned to a wall; in the *Toronto Star*, he was tied to a tree; the *Morning Post* said he had been stuck to a door. As the rumor spread, it morphed from one crucifixion to two, then multiple incidents. The rumors prompted unrest on the streets of London and questions in the House of Commons, including one which added even more baroque details to the crime, alleging that "the Germans had removed the figure of Christ from the large village crucifix and fastened the sergeant while alive on the cross."[142]

Did the crucified Canadian ever exist? There were certainly no substantiated reports at the time, although that didn't stop the allies turning it into fertile propaganda. Subsequent investigations have turned up possible candidates for who the soldier might have been, but none has ever been verified.

But that grim story was nothing compared to "the master hoax" of the First World War: that of the German "corpse factories." Exactly where it began is unclear (it's often claimed to be the work of British intelligence, which it probably was, although that could be a myth in itself), and the details changed regularly. The basic story was always the same: the Germans would transport their dead back from the front lines in corpse bundles, to a factory where their bodies would be processed and boiled down to produce all manner of products—soap, explosives, fertilizer. The factory even had a name: "the great Corpse Exploitation Establishment (Kadaververwertungsanstalt)," as an article in *The Times* had it.[143]

The most plausible source for the story is the British intelligence chief, Brigadier General John Charteris, who, it was reported, had boasted about inventing the tale, at a dinner in New York, in 1925—but he furiously denied the report when he got back to Britain, which could have been because he got told off for running his mouth, or could have been because the report itself was hooey.

The grim tales of the First World War weren't the first atrocity tales to emerge from the fields of war, though—they have a much longer history than that. In April 1782, around the end of the American Revolutionary War, a shocking story appeared in a supplement to the Boston *Independent Chronicle*. It related a horrifying discovery made by one Captain Samuel Gerrish of the New England militia. Eight large packages were seized in transit to the governor of Canada. Upon examination, it was found that the packages contained the most grisly cargo imaginable: a total of almost a thousand human scalps.

Captain Gerrish related how the scalps had been taken, over the course of three years, from their unfortunate American victims by a group of Seneca people, on the orders of the British government. The packages were intended to be sent onward from Canada to no less a person than King George himself, as a gift to lift his spirits.

The newspaper broke down the source of the scalps in grim and unsparing detail. Three hundred and fifty-nine were the scalps of farmers, murdered in their fields or in their houses, eighteen of them specially marked to show they had been burned alive. Forty-three were the scalps of American soldiers, shot dead in skirmishes. Eighty-eight came from women. The article dutifully recorded that a very large number of the scalps had been taken from children: 193 from young boys, 211 from young girls, and, perhaps worst of all, 29 from babies that "were ript out of their Mothers' Bellies."[144]

The story was stunning and scandalous, and other newspapers in cities from London to New York and Philadelphia followed suit in printing their own versions of it over the following months. It caused consternation in Britain, while, in America, passions rose against the British for ordering these awful crimes.

There was just one thing about all this, which you will have guessed already: it wasn't true. None of it. There was no Captain Gerrish, and there certainly weren't any gruesome packages of human scalps winging their way to a bloodthirsty King George.

But not only was the story a fake—so was the entire newspaper. Or, rather, this edition of it was. To be sure, the Boston *Independent Chronicle* was a real publication. To give it its full title, it was the Boston *Independent Chronicle and Universal Advertiser,* because eighteenth-century newspapers were really not big on snappy branding. But the so-called "supplement" was a fabrication from start to finish. The story about the scalps on page 1 was a fiction, as was the letter from the war hero John Paul Jones that followed it, and so, too, were the classified ads for a "large tract of land" and "a convenient tan-yard" that filled out the space at the bottom of page 2.

The whole thing was an extremely convincing forgery, a work of genuine craft and care, and possibly even love. Its deceptive origins would only have been apparent to you if you'd been looking incredibly closely, and also if you were an eighteenth-century typography nerd. If you *were* an eighteenth-century typography nerd, then—in addition to giving thanks that you'd been born two centuries before the invention of Comic Sans— it's possible that you might have noticed the metal type used to print the newspaper wasn't American or English in its style. It was actually French.

That's because the newspaper had been printed, not in Boston, but in Passy, at the time an idyllic, upmarket commune with a rather nice spa, on the outskirts of Paris. Its author had nothing to do with the real *Independent Chronicle*, and he hadn't even lived in America for several years. The person who created the fake paper was none other than the United States' ambassador to France—the Founding Father, polymath, future mesmerism debunker and eighteenth-century typography nerd Benjamin Franklin.

Yeah, him again.

What could have driven Franklin, this man of science and of letters, one of the most revered figures of his age, to come up with such an outrageous deceit? The practical answer is quite simple: it was an act of propaganda against the British. At the time Franklin distributed his *Independent Chronicle* hoax, the Revolutionary War was all but over; the decisive French-

American victory at the Battle of Yorktown was six months past and the Paris peace talks were just starting. Franklin's hoax wasn't actually distributed in America; instead, he sent it to allies in Britain, Spain and the Netherlands. His goal—in which he succeeded—was to seed the tale in the British press, in the hope that it would sway public opinion toward the payment of reparations to America for the cruelties the British had inflicted.

But, while that's the immediate, pragmatic reason, there's also a deeper—and, I think, much more satisfying—answer to why he pulled this con, which is this: Benjamin Franklin really loved lying. Absolutely couldn't get enough of it. From his teenage years until just days before his death at the age of eighty-four, Franklin was the consistent, gleeful perpetrator of wild and extravagant hoaxes, both great and small. Sometimes they were for political purposes, sometimes for financial gain, sometimes for personal pettiness, and often just for the sheer unalloyed joy of making shit up. Others may have misled with greater consequences, but, by any standards, Ben Franklin has to go down as one of the most prolific, skillful and innovative bullshitters in history.

Exactly how seriously Franklin expected people to take most of his hoaxes is…unclear. After all, he was hardly the only person publishing things under fictitious pseudonyms at the time. The rise of the printing press had led to an explosion of what we might call "content," and people were still getting their heads around the fact that some printed things could be true, while others could be made up. Just a few years before Franklin created Silence Dogood in 1722 (for more on this, see chapter 2: Old Fake News), Daniel Defoe had published *Robinson Crusoe*— a strong candidate for the first ever English novel, and one that was written in the style of, and was widely believed to be, a factual autobiography. At around the same time, Jonathan Swift was busy inventing modern satire. Was Franklin's intention to actually deceive, or was he simply experimenting with new literary forms whose ethical boundaries were still a little fuzzy?

The line between "hoax" and "spoof" is a blurry one, even now, and almost all of Franklin's fabrications were united in being outlets for his puckish, satirical and extremely overactive sense of humor. Simply put, Franklin was a leg-puller extraordinaire. According to one possibly apocryphal tale, Thomas Jefferson is said to have explained to people that the reason Franklin wasn't asked to write the Declaration of Independence was "because he could not have refrained from putting a joke into it." I feel I speak for many people when I say that it is a profound shame that history was denied Benjamin Franklin's alternative, funnier Declaration of Independence.

I mean, don't get me wrong, the Declaration as it stands is a solid piece of work, but it's not exactly noted for its banter. It couldn't have hurt to chuck a couple of gags in there to lighten the mood.

But, while many of Franklin's hoaxes were undeniably intended to amuse (himself, at least, and possibly others), that can't really be said of the scalping story. Whatever satirical intent there may be in his writing is far outweighed by anger. And by this point in his career, he was perfectly aware of how widely his tale would be believed, and he knew exactly how to plant a false story in one country's press in such a way that it would spread internationally, copied from paper to paper and passed from country to country. This was a deliberate deceit for devious diplomatic ends. It was done with care: the issue number on the paper was that of the issue published only a month before, it included the name of the paper's real editor and its appearance was a very close replica of the look and style of the real *Independent Chronicle* (although, like many forgers throughout history, Franklin couldn't resist the opportunity to slightly improve on the thing he was supposed to be copying, using an elegant but telltale italic typeface that he'd had custom-made for his own

Passy Press).[145] And, when he mailed the paper to John Adams, he even pulled the old trick of pretending to be skeptical about the fake that he himself had created mere hours before.[146]

Though the target of Franklin's disinformation was the British, the story he told ended up wounding an entirely different set of victims: the Native Americans, about whom he'd constructed what can only be described as a massive racist lie. In his hunt for gory details to give his story a bit of attention-grabbing sensationalism, Franklin repeated, amplified and embellished a falsehood about the indigenous nations that would color perceptions of them for a very long time to come.

To be clear, scalping as an act of war undeniably took place in America, and had almost certainly been practiced by the indigenous people long before the arrival of European colonists. But, from the very beginning of the Revolutionary War, the potent combination of fear, rumor and propaganda blew it up into an ever-present threat, far out of proportion to its actual occurrence. It became a folktale—a universal, constantly lurking boogeyman. Hearsay and whispers meant that false tales of massacres and mass scalpings by Native Americans, supposedly urged on by the British placing bounties on white American scalps, were commonplace—one such hoax even made it into the Declaration of Independence.[147] (It would definitely have been better with jokes instead.) In reality, scalping was by no means the sole preserve of the Native Americans. It was actually practiced by every side during the war, with the revolutionary forces frequently offering large bounties for Native American scalps. Indeed, just a few weeks before Franklin composed his fake newspaper, probably the single worst atrocity of the war had taken place at Gnadenhutten, Ohio, when a white revolutionary militia herded more than ninety unarmed prisoners—Native American men, women and children—into barns, before beating them to death with mallets and then scalping them.

Maybe if Franklin's ploy had stayed where it was intended—in the London press of spring 1782—it would only be a footnote to history today. But, instead, it had an afterlife that extended far beyond the eventual signing of a peace treaty in 1783. Because that's the trouble with really compelling lies: once you put false information out into the world, it doesn't just quietly go away after it's done the job you wanted. Lies are like zombies—they refuse to die, and they're coming for your brains.

That's what happened with Franklin's hoax, which roared back to life with a vengeance more than two decades after his death. In the lead-up to the USA once more fighting the British, in the War of 1812—with some native tribes again siding with Britain—the tale was somehow unexpectedly resurrected. And, this time, its impact was far larger.

When Franklin had first spread his story, eight American newspapers picked it up. On its second go-round, between 1806 and 1814, no fewer than twenty-seven different newspapers published versions of it, twelve of them—ranging from South Carolina to Vermont—in the space of just seven months in 1813. The myth seeped into American public consciousness, adding to the perception of the Native Americans as merciless savages. Despite it eventually becoming public knowledge that Franklin had admitted to the hoax in his letters, the claim still occasionally gets repeated as truth, even in the modern day. We can never know quite how much that outrageous, memorable falsehood contributed to the callous treatment the Native American nations received over the following centuries, but it surely wasn't nothing.

7

FUNNY BUSINESS

W herever there's money to be made, there'll be someone willing to twist the truth to make it.

That's not really surprising. We've built a world where money is important: it lets you live, lets you fulfill your desires, and, if you get enough of it, it lets you have power. When you accumulate enough money and power, you can start to make other people do what you want them to do, and to change the world around you. And, at a certain point, this power grows so large that it can seem like you're able to shape reality to fit your desires. You no longer need to twist the truth to get money; money twists the truth for you. If everybody treats you like the greatest businessman on earth, then, almost by definition, you are. Opportunities open up for you that are closed to others. You can get away with failures that would ruin somebody else. You get to go Full Gatsby, if you want. If Thérèse Humbert thought that money was an illusion—a "conjuring trick that had to be mastered"—then...well, she wasn't entirely wrong.

Basically, you just need to fake it till you make it.

This chapter will look at the various ways that people, throughout history, have faked it—and at least *temporarily* made it—in the pursuit of money.

It's worth noting just how big a role hindsight plays in all of this. In business, the "fake it till you make it" mindset isn't just tolerated—it's regularly taught as a vital lesson in entrepreneurship, along with heroic anecdotes of scrappy upstart chutzpah that get shared among the kind of people whose favorite social network is LinkedIn. For example, Microsoft got its start when Bill Gates, pretending to be his cofounder Paul Allen, called up the president of the company that made the pioneering Altair personal computer and told him that they had written software for it. The president, Ed Roberts, was impressed and asked them to come in and demo it. Which was great...except that Gates's claim wasn't even slightly true. Not only did they not have a finished product—they hadn't even *started* creating it yet; the software was actually written in a frenzy during the two months between the phone call and the demo. Not having an Altair computer to test it on, they had no idea until the day of the demo if their software even worked.[148]

That's hardly the only example of "fake it till you make it" success stories, even if you limit your sample to "globally dominant American tech companies." When Steve Jobs so memorably unveiled the iPhone in 2007, promising a "revolutionary and magical device" that would reinvent the phone as we know it, he had one slight problem: Apple hadn't yet managed to produce an iPhone that actually worked. Their prototypes kept crashing and freezing and dropping calls. When Jobs demonstrated the iPhone, live, to a rapturous crowd at the Moscone Center in San Francisco, seeming to casually switch between apps in a slick display of the iPhone's game-changing power and usability, he was actually following a rigidly defined "golden path"—a precise series of actions, painstakingly worked out by the firm's engineers as pretty much the only sequence the phone could actually get through without breaking.[149]

Of course, the reason why Gates and Jobs have a place on the syllabi of business schools the world over is the simple fact that, having faked it, they subsequently made it. They made judgment calls based on what turned out to be perfectly accurate assessments of their capacity to deliver on promises. The nascent Micro-Soft proved their coding mettle and set themselves up to own the home-computing market; Apple fixed the iPhone's memory issues with a new, custom-made chip and changed forever the way we ignore people on trains. VERY IMPORTANT NOTE FOR LAWYERS READING THIS: I AM NOT SAYING BILL GATES AND STEVE JOBS WERE CROOKS OR CON ARTISTS OR ANYTHING LIKE THAT. They were good at what they did! This book was written in Word on a Macbook Pro! Thank you, lads.

The point is simply that this is all retrospective: you only get to find out after the fact whether history's going to file you under "faker" or "maker." As such, it doesn't provide a terribly useful guide for how to act *now*. "Go ahead and do the naughty thing because, in the future, you will be successful and all this will be little more than an amusing anecdote" is not generally accepted as a legitimate ethical position by any of the world's major religions, other than capitalism, and it very much relies on you becoming successful to justify all the stuff you did before.

The flip side of this, of course, is when people make inaccurate assessments of their capacity to deliver on promises. That's when, instead of getting a Microsoft or an Apple, you get a Theranos—the "unicorn" biotech company with a $10 billion valuation, built on a revolutionary new type of blood test that, it subsequently turned out, couldn't actually test blood very well. This was a canonical example of faking it, not making it, faking it some more, and eventually ending up facing fraud charges. But before their eventual downfall, the company and its young founder, Elizabeth Holmes, were widely celebrated in the media as pioneers, with magazine covers and glowing profiles and comparisons to Steve Jobs—which seemed to be based largely on the fact that Holmes wore black turtleneck sweaters in the hope that people would compare her to Steve Jobs. (Rather than going into more detail here, I'll just tell you to read *Bad Blood: Secrets and Lies in a Silicon Valley Startup*, by John Carreyrou, the journalist who first exposed Theranos in 2015, because, honestly, it's a bit of a jaw-dropper.)

To put it another way, if Gregor MacGregor had just spent a little less time designing imaginary coats of arms and honor systems, and a little more time recruiting people with the skills needed to build a town in the middle of a jungle, would he have even appeared in chapter 5? Maybe he might have managed to pull it off, and now he'd be feted as a heroic pioneer and he'd have things named after him, and books suggesting that actually perhaps he was a bit dodgy would cause diplomatic incidents with the Poyais government, and newspaper columnists would get to write frothingly angry columns condemning university students for petitioning to have statues of him removed.

Exactly how fine those margins are can be seen in the case of Whitaker Wright.

Wright was every inch the model of the superrich late-Victorian industrialist—a self-made man who had raised himself up from poverty, whose business interests spanned continents and whose wealth was shown off in increasingly ostentatious spending. When I say "ostentatious spending," I really do mean that. He had it all: the vast country estate, the mansion in the fanciest part of London, the enormous yacht that he would race against the yacht of his friend Kaiser Wilhelm II. The central feature of his estate, Witley Park, was a series of man-made lakes, one of which housed a showpiece that sounds like something out of a steampunk novel: an underwater smoking room.

This 18-foot-high glass dome with mosaic floors on the bed of the lake, reached via a 350-foot tunnel, allowed Wright's guests to smoke cigars and drink and dance and look up at fishes swimming through the water above. Described by the press of the time as "a submerged fairy room," and by Wright as his "crystal cavern,"[150] smoke was vented out through the mouth of a statue of Neptune that protruded above the surface, and the dome's glass was regularly cleaned by a team of specially hired divers.

Let's say it again: UNDER 👏 WATER 👏 SMOKING 👏 ROOM 👏. If you don't have an underwater smoking room, can you ever really say you've made it?

The thing was, though, Wright's entire business empire was based on lies.

Whitaker Wright's now long-abandoned underwater smoking room. As you might be able to guess from the fashion, this picture is from the '70s.

Wright was British, but he made his fortune first in America, and then in Australia and Canada. He'd first immigrated with his family to Ontario after the printing business that he and his brother had founded in Halifax went bust after a year—an early taste of business failure that Wright seems to have become determined he would never experience again. He relocated to Philadelphia in the early 1870s, when business was booming everywhere. Initially, Wright's financial success was largely legitimate, albeit bolstered by a combination of profound self-belief, a charming and loquacious manner, and some invented credentials. Despite having left school at fifteen, Wright took to adding "MA" after his name, and claimed that he'd studied geology at the prestigious Heidelberg University. He seems to have decided, correctly, that the effort barrier to checking those credentials would prove too much.

Wright became obsessed with mining and decided that this was where his future lay. He had a real talent for assessing the quality of ore, and he went west to join the mad scramble for precious metals. He started in the silver-mining town of Leadville, which had seen an influx of thousands of prospectors in the previous years, and whose mines were churning out millions of dollars a year. It was a wild and dangerous place (Oscar Wilde, visiting it a few years later, said he "noted with alarm that the average bicep size here is thicker than my waist"[151]), populated both by those with extravagant new wealth and by many more people who were desperate for a slice of that wealth.

Initially, Wright took the same path as many others: a hardscrabble life on rocky slopes, hunting for a hint of riches. But he soon fell in with a man named George D. Roberts, a notorious mining magnate of dubious ethics, who had himself been on the sucker end of a famous mining scam a decade earlier (the great diamond hoax of 1872, when a pair of prospectors named Philip Arnold and John Slack seeded some barren land in Colorado with uncut diamonds to convince investors that the land contained a vast fortune just begging to be dug up). This general approach seems to have struck both Roberts and then Wright as

a much easier route to financial success than the business of actually mining; rather than going to the trouble of discovering a
rich seam of silver, just convince eager investors that you've discovered a rich seam of silver and gratefully accept their money.
By the time they realize that what they've bought is just a hole
in the ground, you'll have moved on, with all the power and influence and immunity to consequences that money can buy you.

In many ways, Leadville's richest and most readily exploit-
able resource wasn't its silver—it was its people, and their hunger for a quick buck.

Wright immediately hit on a formula that would serve him
well throughout his life: use his natural charm to convince somebody with unimpeachable credentials and high social status to
join his project and use their reputation and his own imposing
personality to create the impression—especially in an easily ma-
nipulated media—that backing his ventures was a sure thing. His
first human shield of this type was the eminent American paleontologist Edward Drinker Cope, a man who knew an awful lot
about digging things out of the ground, but slightly less about the
human capacity for mendacity. Backed by Cope, Wright managed to float his first mining company on the stock exchanges
for $5 million—considerably more than it was actually worth.

This was the pattern Wright would continue to follow, moving from venture to venture, almost always untroubled by the
fact that he'd left a string of disappointed investors in his wake.
He had a pliant media in his pocket, whose descriptions of him
sounded like something from a romance novel: "A big, burly,
masterful man; everything in his personality breathed pluck, energy, ambition," wrote the *Albany Review*, wiping perspiration
from its fevered brow.[152] When Wright returned from the US
to England to found the London and Globe Company, which
would take his talent for promoting dubious mining stocks to
new levels, he made sure that the board was filled with the great
and the good of London society, including multiple peers of the

realm. Everything was directed toward making sure social pressures kept anybody from asking too many questions.

Ironically, Wright's eventual downfall didn't come from somebody rumbling the fact that his mines weren't always that great at being mines. It came from an entirely different tunneling venture: Wright decided to build a subway line.

Realizing that, with a house built at least somewhat on sand, he needed to diversify his portfolio, and reckoning that something that showed a degree of public spirit could only help bolster his credentials, in 1900 Wright turned to the stalled construction of the Baker Street and Waterloo Underground Railway. This was an underground railway that had been under troubled construction since the early 1890s, having repeatedly hit delays and financial difficulties. (I know, an underground train line in London running over budget and not being finished on time—it's hard to believe.) Wright decided that he could help and floated a bond to finance its ongoing construction.

It was a shitshow. Almost nobody wanted to buy it, and Wright wasn't able to do anything to speed up the construction of the line. The resulting financial pressures on the rest of Wright's business empire led him, in a desperate attempt to maintain the facade that everything was rosy, to move money back and forth in a complex series of loans between his companies, artificially inflating their accounts. But they couldn't withstand the scrutiny this brought; Wright was arrested and tried for fraud. In 1902, he was convicted. He committed suicide in the court's holding cell, just minutes after the verdict was handed down.

The same year, a new company took over the construction of the Baker Street and Waterloo Underground Railway. They quickly finished it, and it's still there today: it is, of course, the Bakerloo line.

Whitaker Wright may have been one of the more ostentatious examples of the thin line between business success and outright swindling, but the kind of investment bubbles that he so gleefully exploited have been common throughout history.

In London in 1720, there was such a frenzy of interest in investing in the South Sea that one group of chancers managed to sell stock described as "a company for carrying out an undertaking of great advantage, but nobody to know what it is."[153]

Months later, in Paris, prisoners were being let out of jail, married off to prostitutes and shipped en masse to Louisiana, on the grounds that they desperately needed settlers there because everybody had a lot of money resting on the fact that the place was an economic miracle in waiting.

Wright also wasn't the first to get into the railway business on dubious grounds. In the 1840s, luminaries as diverse as Charles Darwin, the Brontë sisters and William Makepeace Thackeray poured their money into spurious new railways, with the British Parliament passing a new act almost every week, setting up new parts of a rail network that would never be—and, in some cases, physically could never have been—built.

It's easy to assume that these kinds of sharp business practices are recent inventions, and that dealing with the hassle that dodgy businesses cause is something that people in the past didn't have to deal with. We might believe that spending seemingly half our lives complaining about a crappy business to an uncaring customer-care representative is a new phenomenon, and that, when we moan at a company on Twitter, we're experiencing "first-world problems." It can seem that businesspeople shamelessly lying is fundamentally *modern*.

But, if you've ever thought that, then please meet Ea-nasir.

Ea-nasir was basically Whitaker Wright, except he lived 3,500 years earlier. A merchant who lived around 1750 BC in the city of Ur, one of the great city-states of ancient Mesopotamia (in what, today, is southern Iraq), he seems to have been something of a wheeler-dealer who would trade in everything from real estate to secondhand clothes if he could turn a profit from it. But the bulk of his business appears to have been importing copper from the major trading center of Dilmun, some distance further along the Persian Gulf.

All of us have things we're ashamed of in our lives, but we'd like to think that, if we're remembered at all, it'll be for the good things we've done. Ea-nasir represents a bit of a body blow to that hope. You see, almost four thousand years after he walked the earth, Ea-nasir's name is one of relatively few from that time that lives on in the annals of history—but the main thing that we know about Ea-nasir is that he was a massive bullshitter and a bloody awful copper merchant.

We know this because archaeologists have excavated Ea-nasir's house, where he very helpfully archived all of his correspondence—clay tablets bearing messages from his customers that were communicated to him by professional intermediaries. Ea-nasir seems to have done pretty well at business to begin with, including a good deal of trading on behalf of the king.

But as time progresses, a certain theme begins to emerge from the messages people were sending Ea-nasir. To roughly paraphrase, the theme is "What have you done with my money, you unscrupulous bastard?"

Those clay tablets, with their Sumerian cuneiform script—one of the first written languages in human history—are the world's oldest known customer-service complaints.

The four main complainers that we know of are gentlemen named Nanni, Abituram, Appa and Imqui-Sin. From the context of their messages, they seem to have been financial backers of Ea-nasir's business trips to Dilmun, with the promise that they would get a delivery of high-grade copper ingots in return.

Nanni seems to have been quite miffed that, when he sent his intermediaries to pick up the goods, Ea-nasir instead offered them a load of shitty, low-grade copper, then pulled the old, "Take it or leave it, mate," gambit and refused to refund Nanni's money.

It's worth reading one of Nanni's complaint messages in full, because (despite the quirks of translating directly from Sumerian) it really gives a sense of how little has changed in the last four thousand years when it comes to haranguing dodgy businesses:

> *Tell Ea-nasir: Nanni sends the following message:*
>
> *When you came, you said to me as follows: "I will give Gimil-Sin (when he comes) fine quality copper ingots." You left then but you did not do what you promised me. You put ingots which were not good before my messenger (Sit-Sin) and said: "If you want to take them, take them; if you do not want to take them, go away!"*
>
> *What do you take me for, that you treat somebody like me with such contempt? I have sent as messengers gentlemen like ourselves to collect the bag with my money (deposited with you) but you have treated me with contempt by sending them back to me empty-handed several times, and that through enemy territory. Is there anyone among the merchants who trade with Dilmun who has treated me in this way? You alone treat my messenger with contempt! On account of that one (trifling) mina of silver which I owe you, you feel free to speak in such a way, while I have given to the palace on your behalf 1,080 pounds of copper, and Sumi-abum has likewise given 1,080 pounds of copper, apart from what we both have had written on a sealed tablet to be kept in the temple of Samas.*
>
> *How have you treated me for that copper? You have withheld my money bag from me in enemy territory; it is now up to you to restore (my money) to me in full.*
>
> *Take cognizance that (from now on) I will not accept here any copper from you that is not of fine quality. I shall (from now on) select and take the ingots individually in my own yard, and I shall exercise against you my right of rejection because you have treated me with contempt.*[154]

Sadly, we don't have any record of what Ea-nasir's replies to these complaints were—he kept only his incoming correspondence, and none of his responses have made it into the archaeological record. Were they the Sumerian equivalent of "Your feedback means a lot to us; we apologize if our service has fallen below your expectations"? Or were they the Sumerian equivalent of "Ha-ha! Sucks to be you, loser"?

We can't know for sure, but we can probably get a sense of it from the subsequent messages from Abituram and Appa. Abituram seems to have had a bit of an advantage over Nanni, in that he has some extra leverage with Ea-nasir—one of his first gambits in his messages is to threaten to call in Ea-nasir's mortgages if he doesn't hand over the goods to their intermediary, a chap who goes by the name of Nigga-Nanna.

"The silver and its profit give to Nigga-Nanna," begins Abituram's first letter, continuing with his threat, "Why have you not given the copper? If you do not give it, I will bring in your pledges."[155]

His next letter continues in much the same vein: "Why have you not given the copper to Nigga-Nanna?" it begins, before ending (in case the message hadn't quite got through), "The copper…give it to Nigga-Nanna."

Then Appa joins in the fray: "The copper of mine, give it to Nigga-Nanna—good copper in order that my heart shall not be troubled."

After that, Imqui-Sin joins in, and takes up the refrain: "Give good copper to Nigga-Nanna under seal," he writes, before emphasizing it again, just so that the point is crystal clear: "In order that your heart shall not be troubled give good copper to him."

He then adds—in a plaintive voice that anybody who has been stuck on hold with their broadband provider for two hours will instantly recognize, despite the millennia that separate us from the merchants of Ur—"Do you not know how tired I am?"

Did this all pay off for Ea-nasir? It seems that, much like Wright, it worked brilliantly for a time, but then came crashing down.

As we've already mentioned, we have all these tablets because Ea-nasir's house got dug up by archaeologists in the 1950s. And Leonard Woolley, the lead archaeologist, noticed something interesting about our boy's abode. It's a large, fancy house, as befits his status as Kind of a Big Deal. But near the end of the period in which Ea-nasir was trading, the majority of it appears to have been abruptly annexed and incorporated into Ea-nasir's neighbor's house.

Woolley's conclusion? Our friend was rather suddenly and urgently forced to downsize his home and adopt a somewhat more frugal standard of living—after, presumably, years of being consistently roasted by reviewers on the ancient Mesopotamian version of Yelp finally caught up with him.

Ea-nasir's story tells us that there's very little new under the sun. For as long as we've had civilization, we've also had chancers looking to get a leg up on everybody else. For as long as we've had money, we've had people whose main skill is persuading unfortunate people to part with it. And for as long as we've had writing, we've been sending pissed-off letters to ask where the hell our copper is and why exactly hasn't it been given to Nigga-Nanna.

Of course, you can't talk about deceptive business practices without mentioning the king of turning a buck from fakery:

P. T. Barnum, the *Greatest Showman on Earth*. Barnum is notorious not just for the circuses and novelty exhibits that made him a rich man in the middle of the nineteenth century, but for his—how to put this?—complicated relationship with the truth. Barnum was a notorious faker. His shows regularly featured outright hoaxes such as "the Feejee Mermaid" (the top half of a monkey sewn to the bottom half of a fish, which wasn't even from Fiji—it's thought to have been created in Japan before eventually being sold to Barnum[156]), and the "Cardiff Giant," which wasn't even his own hoax. A supposed ten-foot-tall "petrified man" that had been dug up in Cardiff, New York, the giant was actually a sculpture that had been deliberately planted by a pair of cousins named George Hull and William Newell so it could be "discovered." After it was publicly exhibited to huge profits, Barnum asked to buy it; when he was turned down, he simply created his own fake version of the fake giant, and loudly proclaimed that the other one was the fake.

At the same time, Barnum was also a noted debunker of other fakes, particularly railing against mediums and spiritualists. He even wrote a book about it in 1865, *The Humbugs of the World*, which was an account of "delusions ... quackeries, deceits and deceivers" throughout history. Good idea for a book, if you ask me.

But let's just focus on how Barnum, then a grocery shop employee, first got his start in the entertainment industry with a show that debuted at Niblo's Garden in New York on August 10, 1835—just two weeks before Richard Adams Locke would capture the world's attention with his Moon Hoax. This show was of a single exhibit: a woman named Joice Heth, who Barnum claimed was 161 years old, and had been nursemaid to the infant George Washington. He billed her as "the most astonishing and interesting curiosity in the world!"

Joice Heth, of course, had no connection with Washington and was not 161 years old—she was in her late seventies, decrepit and blind. She was also a slave. It had recently been made illegal to own slaves in New York; Barnum got round this by simply "renting" her.

The exhibition was a hit, fueled by eager, sensation-hungry coverage from the penny press. After the success in New York, Barnum took Heth on tour across New England, attracting crowds everywhere desperate to see and touch this living relic. The schedule was punishing, and too much for Joice's frail body; she died in February 1836. But Barnum was not done with her yet. Urged on by the newspapers and responding to skepticism about her age, he arranged for her body to be publicly autopsied, selling tickets at fifty cents each to an audience of 1,500, who watched as Joice Heth's body was sliced open in front of them.[157]

The surgeon who performed the autopsy declared that it had been a hoax—Joice Heth was no older than eighty. But this news didn't harm Barnum's career; it made him. He played it for publicity, telling story after story to the New York press over the course of many months—Heth was actually still alive, the corpse had been a different woman, Barnum had been hoaxed, Barnum was the hoaxer. He understood how the new media worked better than almost anyone (at one point he was said to have twenty-six journalists on his payroll), and quickly realized that truth and trustworthiness weren't the keys to his career; notoriety and the ability to supply an entertaining story were.

Weirdly, all this is left out of that inexplicably successful Hugh Jackman musical about Barnum's life.

Barnum was also able to play off the obsessions of the age— it was a time of fascination with medical science, from genuine breakthroughs to politically motivated racial theories and, of course, quack medicine.

That's not surprising: if you want to make money and don't have many scruples about how, then there are very few better businesses to get into than the business of curing people's ailments. The history of medicine is filled with snake-oil salesmen peddling fictitious cures and dubious remedies. Quite literally "snake oil" on some occasions—as with the case of Clark Stanley's Snake Oil Liniment, which, in a landmark 1916 case, was tested by the US Bureau of Chemistry and was found to contain precisely 0 percent snake. Mr. Stanley accepted a twenty-dollar fine and bequeathed a whole new phrase for "huckster" to the English language.

Stanley was hardly alone, however. For example, take Hadacol, the notorious and foul-smelling brown liquid that was marketed in the forties and fifties as a cure for a wide variety of ailments—when all it actually contained was a few vitamins and, probably more relevant to its sales figures, 12 percent alcohol.

But the business of fake medicine went a lot further (and got a lot weirder) than simply tricking the gullible into purchasing bottles of liquid that wouldn't cure them of anything. Among the most notable—and successful—quacks in twentieth-century history was "Doctor" John R. Brinkley, better known to his contemporaries as "the goat-gland doctor."

Why was he called the goat-gland doctor? Well. Interesting question. Glad you asked.

It's because he transplanted goat testicles into humans. To help cure impotence.

John R. Brinkley, with radio, probably thinking about goat glands.

Brinkley was not a doctor, despite using the title. He never graduated from any recognized medical school, although that wasn't for lack of trying. He had always felt a calling to the medical profession and enrolled in several medical degrees, but financial circumstances and a, uh, complicated personal life (at one point, he kidnapped his own daughter and took her to Canada) meant that he never finished any of them.

But he wasn't going to let a little thing like that stand in the way of his dream of practicing medicine, so he set up a surgery in Kansas. Initially, this was, in fact, a pretty decent surgery—he did sterling work in treating victims of the 1918 flu pandemic. But it wasn't long before he came up with the notion that would make him rich. Confronted with a patient worried about sexual performance, he somehow ended up transplanting a pair of goat testicles into the man's scrotum.

This, obviously, didn't actually do anything, other than give the gentleman a temporary extra pair of balls (they would gradually vanish over time) and a renewed sense of self-confidence. But the treatment became wildly popular, fueled by Brinkley's aggressive marketing and the fortuitous (for him) coincidence that the wife of one of his early patients subsequently conceived a child. The child was human, for what it's worth, not some kind of nightmarish goat–human hybrid.

Brinkley may not have been a good doctor, but he was an excellent and pioneering advertising man. He was one of the first to realize the potential of the relatively new medium of radio. He set up his own radio station in Kansas, which he used to bullishly tout his treatments. After the Federal Radio Commission eventually took away his radio license (shortly after the Kansas Medical Board stripped him of his medical license), he simply moved over the border to Mexico and continued to broadcast from there, using a high-powered transmitter.

As a result of those two decisions, he decided to get into the world of politics and came within a hair's breadth of becoming the governor of Kansas. Standing for election in 1930 as a write-in candidate, he very likely would have won if it hadn't been for a rule that specified his name only counted if it was written in one specific way, disqualifying tens of thousands of ballots cast in his favor.

Oh, also, lots of his patients died, because he wasn't a very good doctor. It all went wrong for him when he decided to sue a critic of his goat-gland transplanting ways for libel; he lost, and lost badly, and in the ensuing blizzard of lawsuits that came his way—many for wrongful death—he declared bankruptcy.[158] Many quacks may have made their fortunes over the years, but at least one of them actually—if accidentally—made a legitimate medical discovery, and gave our language a word for it. That was Dr. Anton Mesmer, who in the eighteenth century became the talk of first Vienna and then Paris for his innovative medical theories and the remarkable effect they had on members of high society.

Mesmer was promoting his theory of "animal magnetism"— the belief that an invisible, all-permeating fluid supposedly suffused the universe, connecting all living things with the celestial bodies of the heavens. With this theory underpinning his medical work, he would treat people in special consultations, which normally involved staring at them, rubbing them and making them hold an iron rod. This produced an incredible effect on the patient, and the well-to-do elites of Europe flocked to receive Mesmer's treatments.

This all came to head in Paris in 1778.

The king, Louis XVI, was much troubled, as his wife—the notorious cake-promoting influencer Marie Antoinette, some years before she would become embroiled in Jeanne de Valois-Saint-Rémy's necklace scam—had become devoted to mesmerism, acting as Mesmer's patron and drumming up business for him, the aristocracy flocking to his clinics so they could touch themselves with iron rods. So Louis did what any enlightened king would do: he assembled a commission of the finest minds in France to investigate the truth of Mesmer's theory. They concluded that Mesmer's theories were bunk, and the Austrian was run out of town—but that didn't stop his theories living on after him, and enjoying a resurgence in America many decades later.

Of course, what Mesmer had discovered wasn't a theory of biological magnetism, or anything like that. What he'd actually stumbled upon—the reason his treatments seemed to have such a profound effect, and the reason that we still talk about being "mesmerized" today—was, in fact, hypnotism.

What he'd discovered was that the human mind is a very strange thing, and that it can easily be fooled—and, indeed, can often fool itself, even to the extent of producing physical effects.

Which, funnily enough, is what the next chapter's about.

ORDINARY

POPULAR

DELUSIONS

In the country of the United Kingdom, in the run-up to Christmas in the year 2018, a major transport hub named Gatwick Airport was shut down for three days because people saw lights in the sky. This was, of course, hugely disruptive to the people of that country: the second largest of London's airports, closed entirely during possibly the busiest traveling time of the year, because someone was flying a drone around it. About a thousand flights were canceled, and 140,000 people were left stranded. For a troubled nation—riven by internal political disputes and facing the most profound upheaval on the international stage in most people's living memory—it could not have come at a worse time.

It's possible that some readers may have heard about this before. "Wow, gosh, thanks for that," you may be thinking. "What a fascinating and obscure little nugget you've dug up there from the previously untouched fields of ancient history."

In retrospect, the events of those three days seem like something out of…well, an airport thriller. The Gatwick drone was a ghost story for our techno-panic age. Whoever was behind the drone seemed to have an uncanny, almost supernatural knowledge of what was happening; like a movie supervillain, they always managed to stay one step ahead. Every time the drone was spotted, it would vanish again before the authorities could trace it; every time the airport was close to reopening, as if by magic, those eerie lights in the sky would reappear at the last minute. The press was full of tales of the drone's almost taunting behavior, buzzing the control tower before disappearing again; hundreds of people saw the drone, and yet, even in an era when everybody carries an internet-connected high-definition camera in their pocket, it managed to evade being captured on anything more than a couple of completely unverified videos showing a tiny, indistinct gray blob against a big gray sky.

The reason I bring this up is that, several months after the incident was over, and with police no closer to catching a culprit, Chris Woodroofe—Gatwick's chief operating officer—gave an interview to the BBC. In doing so, he was very keen to push back on what the BBC snootily described as "a theory, circulating online, that there had been no drone at all."[159] (The BBC report oddly glosses over the fact that the original source for this "online theory" was…a BBC report of an official police statement about how it was "a possibility that there may not have been any genuine drone activity in the first place."[160])

The evidence for there being a drone was that, in the BBC's words, police had recorded "130 separate credible drone sightings by a total of 115 people, all but six of whom were professionals, including police officers, security personnel, air traffic control staff and pilots." These were people he trusted, Mr.

Woodroofe said: "They knew they'd seen a drone. I know they saw a drone."[161]

Hmm.

My purpose, here, isn't to go, "There definitely wasn't a Gatwick drone, you guys." There probably was a drone! Drones are very common and this totally sounds like something somebody would do! It was probably a hostile nation state doing a test run for something indefinably bad, or an antidrone equipment manufacturer looking to drum up a little business, or just some prick.

I want to be very, very clear about that, because otherwise I can pretty much guarantee that, the day this thing gets published, every newspaper in the land will have the headline, *GATWICK DRONE SUSPECT ARRESTED, CONFESSES EVERYTHING, ALSO PROVIDES DETAILED EVIDENCE DEMONSTRATING THAT THERE DEFINITELY WAS A DRONE.* This is a book, after all, and, unlike the internet, you can't go back and delete the embarrassing bits later. Which, let me tell you, *sucks.*

And yet…a little nagging suspicion remains every time I read that explanation of why there *must* have been a drone. And it's simply because one of our most consistent mistakes as humans is how much we overrate the reliability of eyewitnesses—and our belief that "more eyewitnesses" equals "more reliable eyewitnesses" is, to put it mildly, not necessarily the case.

In the "field indefinite" of untruth that our favorite sixteenth century French essayist Michel de Montaigne talked about, we get misled by many people, as we've seen already in this book. The media deceives us, maps lie to us, con artists fool us, politicians spin to us, businesspeople rip us off and quacks kill us. But the really deep-seated falsehoods? They're not something done to us; they're the ones we do to ourselves. And that nagging suspicion is all the greater because, well, the drone thing all seems a bit familiar.

In the country of the United Kingdom, in spring of the year 1913, people saw lights in the sky. This was a troubled nation, on the verge of profound international upheaval and panicking about new technology. This was the Great Phantom Airship Panic of 1913.

A German airship from 1912.

For several months in the spring of 1913, reports came in from right across Britain and Ireland of mysterious aircraft maneuvering in the skies above the country. There were hundreds of sightings, from thousands of eyewitnesses, from every extremity of the British and Irish isles. Airship reports ranged from Dover in Kent to Bovisand Bay in Devon; from Sanday in Orkney to Galway in Ireland, and everywhere in between; sightings came in from Kirkcaldy, Leeds, London and Portsmouth, from Hornsea, Caernarvon, Cromer, Shepton Mallet, Ilfracombe and Chatham, and from many, many more places.[162]

It actually started in the winter of 1912. At the time, the UK was gripped in a state of profound generalized anxiety. War, everybody could tell, was looming ominously. The people—notably some sections of the press—who were most convinced that war was coming were also the people doing the most to push the country toward war.

Francis Hirst, then the editor of the *Economist*, wrote in his book *The Six Panics*, later in 1913, "In a few days the *Daily Mail* was able to announce: 'It is now established beyond all question that the airships of some foreign Power, presumably German, are making regular and systematic flights over this country.'"[163]

It wasn't, in fact, established beyond all question. The Germans did have at least one airship at the time, but all the historical records suggest it never made it anywhere near the UK. It certainly didn't make hundreds of trips to all corners of the British and Irish Isles over the course of several months. It's possible that one or two of the sightings may have been of an actual aircraft; this was still near the beginning of the aviation age, and experimental flights were not unknown, from either nations or hobbyists. But the vast majority of them—all those lights in the sky, seen by all those thousands of eyewitnesses—can't have been anything other than a huge, nationwide mass hallucination.

One of the fun things about the Phantom Airship Panic of 1913 is that it wasn't even the only one of the time. There had been a smaller precursor airship panic in the UK in 1909, and, in that same year, there was a similar mass hallucination of a flying machine in the United States.

That came about because a chap with the excellent name of Wallace Tillinghast told the *Boston Herald* that he'd invented "the world's first reliable heavier-than-air flying machine" and had flown it three hundred miles from Worcester, Massachusetts, to New York to Boston and back. The fact that nobody had spotted this supposed flight, including the bit where it circled around the Statue of Liberty, was put down to the fact that it had happened at night. (Tillinghast refused to show anybody his vehicle in daylight.)

The deep implausibility of all this didn't stop the people of

New England from reporting a huge number of sightings of the Tillinghast contraption in the skies over the following weeks, all of them eagerly reported by the press. It started on December 20 with a man who said he's seen lights flying over Boston Harbor, which the *Boston Globe* put on its front page under the headline, "Unknown Airship Makes a Flight at Night." The following day's correction—that what the man had seen turned out not to be an airship, but a ship, which turned out to have been in the water rather than in the sky—was slightly less prominently placed on page 12.

By December 22, over two thousand people in the Worcester area had reported seeing lights circling overhead. The next day, further reports of the airship, spread by telephone, brought an estimated fifty thousand residents of Worcester pouring into the streets. On Christmas Eve, there were thirty-three separate sightings in locations as far afield as New York, Vermont and Rhode Island. Mr. Tillinghast's remarkable vehicle certainly had quite the range.

These sightings weren't simply of lights—many of the eyewitnesses insisted that they could see the structure of the aircraft, and could even make out two men sitting in it. Tillinghast himself absolutely played up to it all, acting very mysteriously and disappearing for long stretches of time, before coming back looking windswept.

But then, much as with the Mad Gasser of Mattoon from a few chapters back, the mood of the press suddenly switched. By Christmas Day, one newspaper was openly calling it a delusion, writing that "the epidemic of infected vision that has turned Massachusetts upside struck town with a bang late yesterday." A few days later, the whole press had turned on a dime and were cheerfully mocking the credulous for buying into the hoax.

You'll be astonished to learn that no evidence has ever emerged that Tillinghast actually had a plane.

What all this illustrates is simply that—despite the huge

amount of weight we accord to eyewitnesses, personal experience and multiple reports of the same thing—none of them are necessarily that reliable. We are incredibly prone to deceiving ourselves: we are fallible, suggestible and afraid to go against the crowd. It's a bullshit feedback loop on a societal scale; as every new report adds more heft to the idea that something *must* be true, people start to perform those acts of "spontaneous mendacity," and nobody is willing to admit that maybe there was nothing there all along.

This doesn't just limit itself to false belief—bizarre group manias are commonplace throughout history. It's in the field of rumor and group delusion that our tendency toward infectious irrationality manifests itself most strongly, and this is particularly the case when the thing everybody's imagining is something we can all be scared of. The idea of "moral panics" is a fairly recent invention, but they have existed for many centuries. And the ways they happened in the past seem eerily similar to what we see today.

These have surprisingly common repeated themes: for example, the sudden and widespread fear that some malign external force is causing people's genitals to shrink or even disappear. This is something that's been reported many times throughout history, across many different cultures (the technical medical term for it is "koro"). In 1967 there was an outbreak of shrinking-penis panic in Singapore, with one hospital reporting that at the peak they were treating seventy-five men a day who were convinced that their dicks were becoming smaller and would eventually vanish entirely. In Nigeria in 1990, there was an epidemic of vanishing penises, attributed to sorcery.[164] In medieval Europe, the fear of witches stealing penises (and occasionally putting them in trees) was common.[165]

Humans are weird.

Another major category is panic around contaminated food and drink. Examples of this are common: from persistent rumors that children's sweets have been poisoned, to the rumors that swept parts of the Middle East in the late nineties that chewing gum had been spiked with aphrodisiacs that would cause uncontrollable sexual behavior.[166] But none can quite come close to how consequential the poisoning panic that gripped France seven hundred years ago was.

Right now, there are (entirely justifiable) concerns about the role of technology platforms like WhatsApp in spreading rumors and inciting violence in numerous countries around the world. But it's easy to fall into the trap of believing that, just because a new technology is being used in an event, it must have *caused* the event.

In April 1321, panic spread through the town of Périgueux in southwestern France. Rumor had it that a plot had been uncovered to poison the town's well. Now, in medieval times, poisoning the water supply was pretty much the closest you could get to a weapon of mass destruction.

To understand the context, the continent had been ravaged in the preceding years by a great famine that had killed an unprecedented number. Death was in the air, and people were understandably anxious.

In Périgueux, the gossip mill settled on who was to blame: lepers. The mayor had every leprosy sufferer in the town arrested. Ten days later, each one of them was burned at the stake, their property appropriated and sold to local lords.

But the panic didn't stop in Périgueux. In the following days, lepers were accused of well-poisoning in Martel to the east of Périgueux; in Lisle-sur-Tarn, over a hundred miles southeast; and in Pamiers, almost two hundred miles south.[167]

The theory was that the lepers were trying to spread their disease to the uninfected majority; this was paranoia about disease, yes, but also about conversion and demographic replacement. The lepers "were plotting against the health of the public," wrote the inquisitor Bernard Gui, "so the healthy people, by drinking or using that water, would be infected in such a way, that they could become lepers or die, or would be destroyed from the inside; and in this manner the number of lepers will increase and of the healthy people will decline."[168]

For the next three months, the panic spread from the Toulouse area across much of France, passed from person to person, town to town. It expanded across borders, into modern-day Spain—in early June, King James II of Aragon, fearing that his land was being infiltrated by a foreign threat, ordered a total and complete shutdown of lepers entering the kingdom, until they could figure out what was going on. By late June, he decided that wasn't enough, and had all foreigners arrested.

By July, the king of France had issued orders demanding the arrest and torture of lepers. Hundreds of lepers were killed. In Toulouse, the official accounts for that year added a whole new section just for income generated by confiscating the property of executed lepers.

But then, over the course of the summer of 1321, the conspiracy theory mutated. What had begun as straightforward disease paranoia, spread through person-to-person gossip, now entered the political realm—and, once a rumor becomes political, powerful people will attempt to shape it to fit their agendas. The most notable change that summer was over who was to blame. Suddenly, it wasn't lepers who were at fault; it was Jews, or Muslims.

In Chinon, 160 Jews were put to death.

Eventually, as the summer drew to an end, people finally began to suspect that, in fact, nobody had actually been poisoning the wells at all. King Philip of France eventually ordered that imprisoned lepers should be released, which was probably not much comfort to everybody who had been executed.

But the rumor didn't die there. The immediate panic was over, but the idea that people were poisoning wells continued to spread out across Europe, sometimes going dormant for years before reappearing again. It came back with a vengeance across much of Europe in 1348, as the continent was being ravaged by bubonic plague and infection paranoia reigned. In the German Empire, hundreds of Jewish settlements were burned to the ground as a result.

It's important to remember that this isn't just before the age of social media—it's before the age of mass media, full stop. It's almost three hundred years before the first newspaper. The fastest information could travel was the speed of a horse.

But, nonetheless, lots of elements seem...familiar. It's a rumor passed around, based on nothing at all—fake news, if you will. The idea spreads like wildfire. It goes viral, you could say. It crosses borders, it mutates over time, just cropping up again and again, and it has horrible consequences. Panics about some sinister outside force tampering with our food and drink, in particular, have recurred over and over throughout history, and are still one of the bigger categories of viral rumors on Facebook right now.

Of course, if you're going to talk about these kinds of witch hunts, then you probably need to discuss the most notable witch hunts of all time: namely, actual witch hunts. There are a lot of examples you could choose from to illustrate the witch madness that gripped Europe for several centuries—but let's focus on a witch-hunting king.

By the standards of most historical leaders, James VI of Scotland (a.k.a. James I of England and Ireland) was not actually *that* bad a guy. Moderately sane; managed to hold together a group of countries riven by religious disputes; never seemed to be totally into all the persecution of Catholics that came with his job description; almost certainly got to have a ton of sex with his favorite male courtiers. Bonus: the King James Bible, which is a very nicely written Bible.

But one thing about James is that he was absolutely obsessed with witches.

Not obsessed with witches in an "owned *The Craft* on VHS, dyed his hair black and got really into Neil Gaiman" sort of way. More in a "personally oversaw their torture" way.

James basically introduced the concept of witch hunts to Scotland, kick-starting decades of persecution across the land. Not only did he order Scotland's first large-scale witch trials, he literally wrote the book on it. The book sold pretty well (as you'd hope, given that *he was the damn king*), and it helped stoke a national obsession with witches that would lead to an awful lot of women and quite a few men getting needlessly executed.

James picked up his deadly hobby from the already witch-mad country of Denmark, where he'd gone to collect his new teenage bride, Anne, the sister of the Danish king. Anne's attempt to sail to Scotland upon the arrangement of their marriage had failed due to bad weather, so James went to fetch her and bring her back, only to encounter dangerously stormy conditions himself that left him stuck there. In the end, having left in October 1589, James didn't get back to Scotland until May the following year, during which time the couple married, had a honeymoon, visited the sights, hung out with hard-drinking genius astronomer Tycho Brahe and generally seem to have had a pretty good time.

But James always had a touch of paranoia (not unreasonably, given lots of people genuinely did want to assassinate him), and he returned to Scotland brooding heavily over their frustrated voyages. Pondering the question "How could the weather possibly have been so bad all winter?" James naturally came up with the answer "Because witches," rather than the more obvious "Because you live in bloody Scotland." "Oh, yes," said his new Danish friends, nodding sagely, "that was absolutely witches. Classic witches."

And so began the North Berwick witch trials, in which no fewer than seventy people were tried for various witch-related activities. The main "culprits" were brutally tortured into confessions, with James himself participating in several of the torture sessions. Among the confessed activities recorded in "Newes from Scotland," James's pamphlet broadcasting his witch-hunting triumph, were the following: kissing the buttocks of the Devil; being licked in a private place by the Devil; being carnally used by the Devil; and causing the storms by throwing a cat into the sea.

King James, examining some witches, from his book Daemonologie.

Over the next five or six decades, an estimated 1,500 people in Scotland would be executed for witchcraft. Which is a lot, but it pales in comparison with the German-speaking areas of the Holy Roman Empire, where as many as 25,000 people—mostly women—were put to death. In total, it's possible that 50,000 people died across Europe during the period of witch mania. Which is maybe worth remembering the next time someone suggests that they're the victim of "the greatest witch hunt ever."

Why? What the hell was everybody thinking? Lots of explanations have been put forward, many of them related to the fact that the seventeenth century was a time of extremely YOLO religio-socio-political chaos across the entire continent. (It's sometimes referred to as "the General Crisis" due to the enormous amount of shit that was kicking off all over the place—remember, this was the same era in which the Ottomans were going through their period of, uh, unorthodox leadership.) Was the witch mania due to economic crises? The Little Ice Age? Attempted gendercide? Just a cunning way to get rid of people you didn't like?

(This isn't a joke—it really is a theory by some very respected anthropologists. Short version: examining English witch trials, it appears that most of the accused seem to have been absolute dicks as neighbors, and everyone was heartily glad to see the back of them.)

One recent study even argued that witch mania was adopted and heightened by the rival Catholic and Protestant Churches as, effectively, a sales technique. (As the paper itself puts it: "Europe's witch trials reflected non-price competition between the Catholic and Protestant churches for religious market share in confessionally contested parts of Christendom."[169]) In other words, areas where Catholic and Protestant confessionals were in direct competition saw lots of witch trials; areas where the Catholic Church remained dominant saw very few. I've no way of judging how accurate this theory is, however, because I'm not a witch economist.

But whether you pick one single explanation for Europe's witch mania or decide it was probably a little bit of all of them, it's never going to be a Unified Theory of Witch Hunts. Because, of course, witch hunting isn't just a European pastime. Depending on exactly how you define "witch" and "hunt," you can make the case that witch hunts have happened at some point in history in virtually every culture around the world.

Ultimately, it might well come down to a fairly fundamental problem that humans have: when presented with the mind-boggling complexity of the world and all the frustrations of living in it, we quite like to be able to point our fingers at a group of people who are not us and say, "Their fault!" And, if we don't do it ourselves, there'll normally be somebody else who stands to benefit from telling us who to blame. Witches have been a historically popular choice, but other common scapegoats include recent migrants, Jews, communists, the Illuminati—sometimes all four of those at once, if you're lucky.

This obviously touches on belief at its most fundamental, and it's no surprise that religion can form the backbone of some of our wilder delusions. To choose just one especially baroque example: in 1962, two con-artist brothers wanted to run a scam in Yerba Buena, Mexico. They decided that the population were credulous and could be taken in by a grift involving ancient Inca treasure and returned gods. To this end, they hired an impoverished prostitute named Magdalena Solís from a nearby town and convinced her to play a reincarnated Inca goddess who would lead a cult. Unfortunately, Solís got so into being a goddess that she came to believe it was true, and—as goddesses often do—started demanding blood sacrifices. At least four people were murdered so that Solís and her followers could drink their blood.

We like to think that we've left the monsters of our imaginations in the past, back in the bygone days, before we were modern and fancy, when everything was shadow and darkness. But the monsters made the journey with us; they're always there, it's just that sometimes we give them new faces or different names.

That's why, for example, in the winter of 1929, a posse of twenty armed men set out into the pine woods of New Jersey, looking for an actual monster. The monster in this case was, as you might be able to guess, the Jersey Devil—the notorious piece of local folklore common to the area. The *New York Times* report of this particular monster hunt describes the Jersey Devil as a "mysterious apparition variously described as breathing fire, having wings, bearing tusks, hairy and blood-curdling."[170]

Interest in the Devil of Jersey had been kept alive by a combination of gossip and rumor, fueled by consistent press attention. The monster hunt of 1929 was sparked by two reported sightings—one by a farmer, who found his pig slaughtered and tracked a set of four-toed footprints into the woods; the other by two schoolchildren, who had come across a "shaggy, black monster with a pig's snout, emitting uncanny cries"[171] in the woods one afternoon.

The police were called. Dogs were sent into the woods to try to get the scent of the beast. The posse was formed and went out into the pines, but found nothing.

I mean, of course they found nothing. There's no such thing as the Jersey Devil. But, nonetheless, belief in it and a constant stream of stories has helped keep the Devil alive, with regular sightings throughout the nineteenth and twentieth centuries; no doubt it will carry on into this century, as well. In fact, the legend of the Jersey Devil stretches back even further—according to the local legend, it first came to terrorize the area in 1735, when a local woman named Ma Leeds gave birth to a hideous monster in Burlington, New Jersey. Originally, it wasn't even known as the Jersey Devil; it was known simply as the "Leeds Devil."[172]

Those of you who've read this book in the traditional front-to-back order may notice something familiar about aspects of the previous paragraph.

1735. Burlington, New Jersey. Leeds.

That's right—the Jersey Devil was originally a myth about Titan Leeds's family, and it was born in the very same year that Benjamin Franklin announced Titan's untimely death.

I'd love to be able to tell you, here, that it was Franklin himself that started the legend. That would be an excellent nonfiction flourish and a triumphant finish to this book. Unfortunately, I can't. I mean, he *might* have started it, I guess, but he probably didn't, and there's no evidence either way. History isn't quite that neat, and so you'll have to go into the final chapter utterly bereft of a satisfying Ben Franklin callback. Instead, it's more likely that Franklin and the originators of the legend were simply responding to the same thing—those old calumnies about Daniel Leeds being "Satan's harbinger," the local petty religious infighting and politics that, two centuries later, would see men with guns prowling through the woods, looking for a devil that exists only in their imaginations.

We didn't leave our monsters in the past. They've been with us every step of the way.

CONCLUSION

TOWARD A
TRUTHIER FUTURE

Early in 2018, I stood in the ruins of the Mayan city of Tulum and watched a small, adorable mammal gleefully eating the flesh out of a coconut. The animal in question was a coati, also known as a Brazilian aardvark—a relative of the raccoon, but cuter, and with less of an air of witchcraft about it. I was delighted to spot one of them, because the Brazilian aardvark is an animal that tells us an awful lot about truth, and how bad we are at it.

You see, there's one thing that's particularly interesting about the coati (also known as the Brazilian aardvark). The interesting thing is this: it isn't actually known as the Brazilian aardvark at all. Or, at least, it wasn't until 2008, when things went weird.

That's when Dylan Breves, a student from New York, went on holiday to Brazil, saw some coatis and thought—very wrongly—that they were aardvarks. Not wanting to be embarrassed by his woeful lack of mammal knowledge, he jokingly made a minor edit to the Wikipedia page for the coati, inserting the claim that (you've guessed it) they were also known as the Brazilian aardvark.

As far as we can tell, before that exact moment—11:36 p.m. Brasília time, on July 11, 2008—nobody had ever used the phrase "Brazilian aardvark." It hadn't been written on the internet, it had appeared in no scholarly articles and it had never been printed in a book.[173]

Now, normally, a little light Wikipedia vandalism like that would be quickly caught and removed by the site's ever-vigilant army of volunteer editors. But for whatever reason, despite the fact that aardvarks do not live in South America and literally no one had ever written the phrase "Brazilian aardvark" before Dylan, this one slipped through the net.

And then, because it was on the internet and people trust Wikipedia, it wasn't long before people started calling the coati a "Brazilian aardvark" for real.

As the New Yorker's Eric Randall reported in 2014, by that date newspapers like the Daily Mail, the Telegraph and the Independent had all picked it up and run it uncritically.[174] The BBC had also used it.[175] "Brazilian aardvark on the loose in Marlow," shouted the headline of one local paper in Buckinghamshire when a coati escaped from a private collection. "So that's what an aardvark looks like," runs the headline in another local paper in Worcester, above a picture of a coati not looking like an aardvark.[176] You can find photos of coatis captioned as Brazilian aardvarks on the websites of Time and National Geographic, while Scientific American even went as far as flipping the traditional name order in an article on conservation, calling it a "Brazilian aardvark, also known locally as coati."[177] There now appears to be at least one serious scientific paper from a group of actual Brazilian zoologists which uses the name,[178] and this completely made-up phrase has been used in books from at least two of the world's leading academic publishers. One is from University of Chicago Press ("The coati, also known as the hog-nosed coon, the snookum bear or the Brazilian aardvark"[179]); the other, from Cambridge

University Press, rather wonderfully repeats the mistake in a passage about the great eighteenth-century naturalist, Buffon, criticizing other naturalists for repeating mistakes by copying from other naturalists. "The multiplication of errors was one of the most common features of eighteenth-century natural history."[180] Indeed.

All of this raises the question: Is it even wrong anymore? Is the coati in fact now *actually* also known as the Brazilian aardvark? Did a dumb joke manage to change an animal's name, just because, if something is on Wikipedia, then it will spread out into the world until it becomes sort of true?

The answer, as is often the case, is, "Er, maybe." The Wikipedia page for coatis no longer features the claim that they're also known as the Brazilian aardvark, on the grounds that there's not enough evidence of it being in widespread use. And, since 2014, when the *New Yorker* article came out and the claim got deleted, references to it in the wild do seem to have slowed down somewhat (there was one mention in the *Guardian* in 2017, but that might have been an in-joke[181]). But there's no doubt the Brazilian aardvark is out there now, in the wild, and that, if all of us just agree to start calling coatis by entirely the wrong name, then, damn it, that's what they'll be called.

A Brazilian aardvark, pictured enjoying a snack in Tulum, Mexico.

This might sound like it's intended as a cheap joke at the expense of Wikipedia, which it really isn't—although, in fairness, this is far from the only incident of its type involving the site. There's the regrettable case of the inventor of the modern hair iron, in which a correct reference—Madam C. J. Walker, a pioneering African-American entrepreneur—was replaced in August 2006 with "Erica Feldman (the poopface)."[182] Wikipedia admins quickly noticed the vandalism...and removed only the words "the poopface," leaving Erica Feldman, whoever the hell she is, with the credit. The problem was fixed long ago on Wikipedia, and yet, if you google "Erica Feldman hair straightener" today, you'll still find a vast number of websites that will cheerfully tell you about Ms. Feldman's contributions to African-American hair care.

Oh, and there was also that time the report of the Leveson Inquiry (Lord Justice Leveson's examination of "the culture, practices and ethics of the UK press") named a twenty-five-year-old Californian student named Brett Straub as one of the founders of the *Independent* newspaper, because one of Brett's friends had added his name to Wikipedia as a prank.[183] To say the UK press enjoyed that one would be understating things a little.

In fact, Wikipedia even has a list of times this has happened, under the title "citogenesis"—a term coined by *xkcd* cartoonist Randall Munroe—which includes gems such as "the first commercial cardboard box was produced in England in 1817 by Sir Malcolm Thornhill" (now replicated all over the internet) and an entirely invented disease called "Glucojasinogen," which has subsequently appeared in multiple scientific papers.[184]

Readers with long memories may recall that, near the beginning of chapter 2, I wrote, "I promise that I'm not going to make a habit of cut-and-pasting from Wikipedia in this book."[185] I can only apologize. I lied. Deal with it.

But the thing is, in all of these cases, the problem isn't so much Wikipedia as it is people blindly copying from a single source and assuming it's correct (and further people taking that new source as evidence that the first source was correct, and so on). As we've seen time and time again in this book, this kind of circular reporting isn't something that's limited to the internet age; bullshit feedback loops have been with us since the invention of print, and probably long before. The fact that the naturalist Buffon was complaining about exactly the same thing in the late 1700s should probably tip us off that our issue here possibly isn't Jimmy Wales's excellent invention.

It's quite easy for us to blame Wikipedia (or Twitter, or telephones, or the printing press) for a long-standing systemic problem in the ways we gather and distribute knowledge, because blaming new things is easy and fun. But it does miss the point rather. That's something that was shown up in a cheeky experiment that an Irish student named Shane Fitzgerald conducted in 2009, when news broke that the French composer Maurice Jarre had died. Realizing that the world's journalists would be heading to Jarre's Wikipedia page, Fitzgerald fabricated a too-good-to-not-use quote from the maestro—"When I die there will be a final waltz playing in my head that only I can hear"—and quickly added it to his page. This particular bit of vandalism was caught and deleted rapidly, but in the brief window of its existence the quote still made it into many of the world's leading newspapers. And, unlike Wikipedia, none of them caught and deleted it; that only started to happen a month later, when Fitzgerald wrote to them to tell them what he'd done. By this test, Wikipedia was actually considerably *more* reliable than the world's press.

If anything, Wikipedia—and the internet generally—just lets us lift the lid on the kind of mistakes we've been making for a very long time. Anybody with a data connection can go and see for themselves, down to the minute, the exact moment when the false idea that coatis are called Brazilian aardvarks entered the world. For the pre-internet age, tracking something like that down was the stuff entire PhDs were made of.

This is a real problem with history—there's a lot we don't know, and there's also a lot that we think we know that we might not actually know, except unfortunately we just don't know what we don't actually know. Take, just for example, the story of the incredible coincidence that led to the First World War starting. The assassination of Archduke Franz Ferdinand by Gavrilo Princip in Sarajevo on June 28, 1914, all came down to the fact that Princip just happened to stop and buy a sandwich from Moritz Schiller's delicatessen—a sandwich that he was eating when he saw the Archduke's limousine (which had diverted from its planned route) drive past. He seized his opportunity, and the rest is...well, history. If Princip hadn't felt peckish at that exact moment, or maybe if he'd decided that he wanted something different for lunch, then he would never have been in the position to fire the fateful shot, and perhaps the continent would not have descended into war.

It's a great story about how the tiniest of details can have huge outcomes. It's also completely untrue.

The source of the tale appears to have been a BBC documentary from 2003 that included the sandwich story—although, according to the journalist Mike Dash, who tracked the origin of the sandwich story down, the director of the documentary can't remember where they got the sandwich detail from—and it spread like wildfire. It now appears all over the internet, and was even included in a book by the respected BBC journalist John Simpson, which was titled, er, *Unreliable Sources*.[186]

This isn't a new phenomenon. If you're a fan of financial bubbles, you may have been surprised that, in listing other financial bubbles a few chapters ago, I didn't mention the "tulip mania" of 1637. This was possibly the most famous financial bubble of all time, where the price of tulips in the Netherlands soared in price before collapsing, leaving many tulip speculators ruined.

It's been a staple of discussions about the human tendency for foolishness ever since it appeared in Charles Mackay's classic 1841 book, *Extraordinary Popular Delusions and the Madness of Crowds* (from whom I shamelessly poached the title of chapter 8, and also basically the idea for this book). Unfortunately, it also seems to have been, if not completely false, at least wildly overstated; Mackay got his information from a pamphlet put out by opponents of financial speculation, and, in reality, nobody was ruined by the rise and fall in the price of tulips.

The problem that lots of things we think we know turn out to be resting on shaky foundations isn't limited to history either. Right now, science is going through a "replicability crisis"— where we're discovering that an awful lot of bits of knowledge that we thought were well-founded are actually possibly entirely illusory. This all comes down to one of the foundational bits of the "scientific method" (note to sociologists of science: yes, I know there's no such thing as a singular scientific method—let me live). That's the fact that scientific experiments are set up to let anybody else replicate them—that's why schoolchildren are drilled to write up their attempts to prove Newton right with the classic form of Aims, Methods, Results, Conclusion.

The trouble is, a lot of the time, nobody's actually bothered to replicate the major experiments. That's partly due to the incentive structures in science: nobody gets the big grants or the prestigious university posts on the basis of copying what someone already did before. If you want to get ahead in academia, you need to produce new, original work that expands our knowledge. Which regrettably means that nobody bothered double-checking a lot of what we *thought* was our existing knowledge.

This is particularly acute in the field of psychology, where some recent large-scale efforts to replicate a bunch of highly cited, widely referenced studies have come back with the disturbing conclusion that around 50 percent of them may not actually replicate—they might just have been chance findings all along. What's even more interesting is that, deep down, it seems like experts in the field have an inkling which results are dodgy. The experimenters gave a large group of experts not connected to the study a betting market, where they could place wagers on which experiments they thought would replicate and which wouldn't. The betting markets proved uncannily accurate, which is perhaps good news for fans of the human desire to make a quick buck, but less good for the system of peer review.

Oh, and if anybody's going, "But that's just psychology—it's not even a real science, anyway," then, fun news: there's a replication crisis in physics too. Stick *that* in your pipe and smoke it, Einstein. (For what it's worth, it's now believed that around 20 percent of Einstein's published papers contain mistakes of some kind. A lot of the time, he seems to have somehow come to the right conclusion despite the fact that he was working off incorrect assumptions. That's geniuses for you, I guess.)

So where does all this leave us? Is truth in crisis? Are we doomed to live out our lives in a fog of misinformation? Deep down, are we all little more than coatis, hopping around the ruins of an ancient civilization, with tourists pointing at us and going, "Look, Doris, it's a Brazilian aardvark"?

I think not. For sure, yes, we all swim in a sea of half-truths and sort-of lies, because the world is dumb and complicated, nobody knows exactly what's going on, and that's just the way our brains are made. But that isn't a crisis. That's just how things have always been.

The quote that began this book, from the reckless Arctic explorer Vilhjalmur Stefansson—"The most striking contradiction of our civilization is the fundamental reverence for truth which we profess and the thoroughgoing disregard for it which we practice"—might sound like it's from a work that is going to bemoan our failure to live up to the standards of truth. But, actually, he takes the opposite tack, suggesting that maybe we shouldn't be so surprised by the fact that truth is a bit thin on the ground. "It is a bit naive of the philosophers to diagnose from the mere scarcity of truth that the world is sick with an incurable malady," he writes. "Is it not just possible that they cannot cure us for the basic reason that we are not ill?"[187]

I think this is the first thing we need to do if we want to move the needle back from untruth and toward truth: we need to not freak out. We have to appreciate that bullshit will always be with us, and the best we can ever hope to do is to keep it in check.

But there are also some practical things I think we can do—both as a society, and for ourselves.

We need to counter the effort barrier, and the way to do that is to…well, put in a bit more effort. That means being willing to pay for people to actually check things (I mean, I'm a fact-checker, of course I'm going to say that), but it also means that all the different groups in our society whose job is roughly in the truth field need to get a lot better at working together. Academics need to learn to talk to journalists, journalists need to learn to talk to academics, and ideally, if they could not do this only via the medium of press release, that would be great.

But we can all also help to counter the effort barrier ourselves, simply by putting in a tiny bit of effort the next time you're tempted to share something outrageous on the internet. Just a few seconds. Check the source. Google it. Ponder whether it seems too good to be true.

Speaking of which, we also need to check ourselves—any of us, no matter how committed we think we are to the truth, can easily fall into the ego trap and find ourselves liking the lie. In fact, the more honest we think we are, the less likely we might be to be on the alert for these kinds of biases. So, when you're pausing to check the source of something, also ask yourself if it's playing to your personal biases, and whether you're approaching it as skeptically as you might. And we can reflect this up into wider society—everybody makes mistakes, and we need to get better at celebrating those of us who are open to admitting them. Yes, ideally politicians wouldn't say wrong things in the first place, but, hey—at least let's give them a little bit of credit when they correct themselves.

We're also going to need to fill the information vacuums that exist. That's an ongoing process, of course, carried out around the world every day by millions and millions of people, working in a wide variety of fields, who strive to increase the sum of knowledge by a fraction. But we can still do more: too much information that does exist is locked away, hiding in a database or in an unreleased report or behind a paywall. We have to step up our efforts to make more of that good information available widely, because, without it, bad information will just flow right back in to fill the void. It's not enough for us just to pull up weeds in the information garden—we need to plant flowers, as well.

And we need to believe it will work, and that it matters. Giving up and deciding that nobody cares about truth just because the candidate you preferred lost an election is, shall we say, a little premature. Believing that the internet is just a giant bullshit engine and there's nothing anybody can do to tame it is almost as bad. As this book's shown, this is very far from the first time in history that we've had these worries. Uncontrolled rumors, panics over new communications technology, horror at false news and fears of information overload—they've all been around for centuries. We got through it then, and we can get through it now, just as long as we don't throw our hands up and go all, "LOL—nothing matters." The greatest worry about the idea of "fake news" isn't actually that people believe false news—it's that they stop believing real news.

And we need to celebrate the times when we get it right, because sometimes we really do make big steps forward. And sometimes that happens in the most unlikely places—like, for example, a back garden in Paris.

All the same questions that we've pondered here—how to

disentangle small, unexciting truths from a glittering skein of thrilling nonsense—confronted the upright citizens of that city in the 1780s, when our old friend and quack Dr. Anton Mesmer rolled into town. As we mentioned in chapter 7, King Louis XVI was not hugely pleased that Marie Antoinette was letting Mesmer work his hypnotic charms on her. And so he assembled a sort of Empiricism Avengers to test Mesmer's theories. The group included some of the finest minds in Paris at the time, such as the father of modern chemistry, Antoine Lavoisier, and the renowned doctor Joseph-Ignace Guillotin (who, the following year, would propose an invention Louis XVI would ultimately become very familiar with).

In their pursuit of truth, the members of the commission did something that, as far as we know, nobody had ever done before in scientific history. They conducted the world's first ever placebo-controlled, blinded medical trials. In the back garden of the lead author's house, the commissioners invented a fairly hefty chunk of the scientific method, as—pioneering the concept of a blinded experiment in a very literal way—they led a literally blindfolded subject around and got them to hug supposedly "magnetized" trees (before eventually fainting). Through this and other controlled experiments, they conclusively proved that Mesmer's theories were bunk.

You might think that, when they came to write up their findings, they'd have been tempted to brag about this triumph of truth over bullshit. But, instead, they struck quite a different tone: they were almost celebratory about Mesmer's wrongness, finding it far more fascinating than the mundane truth.

"Perhaps the history of the errors of mankind, all things considered, is more valuable and interesting than that of their discoveries," the report's lead author wrote. Echoing Montaigne's observation from centuries earlier, he continued, "Truth is uniform and narrow; it constantly exists, and does not seem to require so much an active energy, as a passive aptitude of soul in order to encounter it. But error is endlessly diversified; it has no reality, but is the pure and simple creation of the mind that invents it. In this field, the soul has room enough to expand herself, to display all her boundless faculties, and all her beautiful and interesting extravagancies and absurdities."[188]

This book has covered just a tiny fraction of that "history of the errors of mankind." You could write a hundred other versions of the book with no overlap.

Hopefully we have managed to follow in the footsteps of the author of that foundational piece of debunking, poised as he was in the very human state of being caught between the push and pull of fact and fiction—the pioneering truth-seeker who, nonetheless, seems to have been strangely captivated by the inexhaustible, soul-expanding possibilities of untruth. Because that's what we need to do if we're going to become more truthful—we need to study more deeply the vast and bountiful fields of wrongness, to know better what it is we're doing wrong before we try to do it right. Basically, we need to become scholars of bullshit.

Oh, what was the name of that author, the one whose back garden that pioneering piece of truth-seeking was carried out in?

It was, of course, Benjamin Franklin.

★ ★ ★ ★ ★

268

ACKNOWLEDGMENTS

There are a lot of people I need to thank, and apologize to. Foremost among these are Ella Gordon, my wonderful editor at Wildfire, and Antony Topping, my excellent agent. Both have treated me with far more patience, encouragement and lack of shouting than I deserved through this book's (slightly delayed) writing process.

I must also thank my wonderful colleagues at Full Fact, who have been both patient and inspiring. Particular thanks to our director, Will Moy, who both allowed me to write the book and was very good at taking hints about not asking how it was going.

Finally, thank you and moreover deep apologies to all my friends, whom I basically haven't seen for many months. I can come to things now, guys! Please invite me to stuff again.

FURTHER
READING

There are many superb books that I relied on in the writing of this book. Most of them are referenced in the endnotes, but if you're keen to dive further into any of the topics in the book, then here's a quick list of some of the best:

In general

I have to shout out the brilliantly titled *Oxford Handbook of Lying*, which was published during the writing of this book. It's the first cross-disciplinary compendium of the latest academic knowledge on lying, and as such briefly made me wonder if I should just not bother writing this one at all. It's great, and is also heavy enough to serve as a useful doorstop, or possibly murder weapon. My book's got more jokes, though. The equally well-titled *Penguin Book of Lies* (currently out of print) was also useful, and all my classical references are stolen from it.

Chapter 1

For more on how, why, and how often we lie, Robert Feldman's *Liar: The Truth about Lying* is enjoyable. Harry Frankfurt's *On Bullshit* (and his follow-up, *On Truth*) is both short and essential. You'll probably have a hard job tracking down Vilhjalmur Stefansson's *Adventures in Error*, but I quote from it quite a bit so should probably give it a shout-out.

Chapter 2

For fans of the two Benjamin Franklin segments, please go seek out Brian Regal and Frank J. Esposito's *The Secret History of the Jersey Devil*, and Max Hall's *Benjamin Franklin and Polly Baker: The History of a Literary Deception*. If you're interested in the invention of news, I can't recommend anything higher than Andrew Pettegree's *The Invention of News: How the World Came to Know about Itself*. On a more academic note, Brendan Dooley and Sabrina A. Baron's *The Politics of Information in Early Modern Europe* has loads of great stuff.

Chapter 3

On the Great Moon Hoax, Matthew Goodman's *The Sun and the Moon: The Remarkable True Account of Hoaxers, Showmen, Dueling Journalists, and Lunar Man-Bats in Nineteenth-Century New York* is great. For media wrongs more generally, check out Curtis D. MacDougall's classic *Hoaxes*, and Robert Bartholemew's *Panic Attacks: Media Manipulation and Mass Delusion*.

Chapter 4

Edward Brooke-Hitching's *The Phantom Atlas: The Greatest Myths, Lies and Blunders on Maps* came out while I was writing this book, and I was both delighted and annoyed to discover that he'd written about almost everything I was planning to. If you liked this chapter please go and read his book, because it's got loads more, and much prettier maps. Go for Bruce Henderson's *True North: Peary, Cook, and the Race to the Pole* if you enjoyed the arctic material.

Chapter 5

David Sinclair's *The Land That Never Was: Sir Gregor MacGregor and the Most Audacious Fraud in History* has lots more about the Cazique of Poyais. Tamar Frankel's *The Ponzi Scheme Puzzle* is excellent on con artists, as is Amy Reading's *The Mark Inside: A Perfect Swindle, a Cunning Revenge, and a Small History of the Big Con* and Maria Konnikova's *The Confidence Game: The Psychology of the Con and Why We Fall for It Every Time*. Hilary Spurling's *La Grande Thérèse: The Greatest Swindle of the Century* is a delight on Thérèse Humbert's life and times.

Chapter 6

For more political lies, have a look at Adam Macqueen's *The Lies of the Land*. For WWI's falsehoods, see James Hayward's *Myths & Legends of the First World War*.

Chapter 7

Have a read of Henry MacRory's *Ultimate Folly: The Rises and Falls of Whitaker Wright* and R. Alton Lee's *The Bizarre Careers of John R. Brinkley* (there are other Brinkley books, but that's the only one I could find that's available in the UK).

Chapter 8

He's already got one shout-out in this segment, but if you like finding out more about our weird beliefs and manias, then go looking for Robert E. Bartholomew's *Hoaxes, Myths, and Manias: Why We Need Critical Thinking* and *A Colorful History of Popular Delusions*. On that note, Charles Mackay's *Extraordinary Popular Delusions and the Madness of Crowds* is a classic (which is why both Bartholomew and I are riffing off his title). Although some bits of it are wrong.

Conclusion

Don't really have any books to plug here, so I'll just note that you should read my previous book, *Humans: A Brief History of How We F*cked It All Up*, too.

PICTURE CREDITS

p. 34 Niccolò Machiavelli, oil painting by Santi di Tito, second half of 16th century. Photo: Imagno/Getty Images

p. 37 Jonathan Swift, oil painting by Charles Jervas, c.1718. Photo: Granger

p. 62 Benjamin Franklin, oil painting by Joseph-Siffred Duplessis, 1785. Photo: Wim Wiskerke/Alamy

p. 69 Print-seller, woodcut, 1631. Photo: Interfoto/Alamy

p. 89 *New Discoveries on the Moon*, engraving by the Thierry Brothers, published 1835. Photo: SSPL/Getty Images

p. 128 *A map shewing the progress of discovery and improvement, in the geography of North Africa*, published by James Rennell, 1798. Library of Congress, Geography and Map Division

p. 132 Map of Africa published by Aaron Arrowsmith, 1802. Library of Congress, Geography and Map Division

p. 134 Map of Africa from *A New Universal Atlas* published by

S. Augustus Mitchell, 1849. Library of Congress, Geography and Map Division

p. 155 General Gregor MacGregor, mezzotint by Samuel William Reynolds after Simon Jacques Rochard, 1820-35. Photo: The Picture Art Collection/Alamy

p. 157 Dollar bill from the Bank of Poyais, 1820s. Photo: History and Art Collection/Alamy

p. 170 Jeanne de Valois-Saint-Rémy, comtesse de la Motte. Stipple engraving by F. Bonneville, 1796. Photo: API/Gamma-Rapho via Getty Images

p. 180 Thérèse Humbert, c.1903. Photo: Harlingue/Roger Viollet via Getty Images

p. 183 Removal of the Humbert's safe, May 1902. Photo: Hulton Archive/Getty Images

p. 193 Titus Oates, engraving by F. Wentworth, c.1880. Photo: Hulton Archive/Getty Images

p. 198 Rose Mary Woods demonstrating "The Rose Mary Stretch," 1973. Photo: Everett Collection Historical/Alamy

p. 217 Underwater dome and tunnel at Witley Park. Photo: Associated Newspapers/Shutterstock

p. 228 John R. Brinkley. Photo: Underwood and Underwood/The LIFE Images Collection via Getty Images

p. 240 German airship. Photo: Interfoto/Alamy

p. 248 King James examining witches. Photo: Chronicle/Alamy

p. 259 Brazilian aardvark. Photo: Tom Phillips

ABOUT THE
AUTHOR

Tom Phillips is an author and journalist from London. He is currently the editor of Full Fact, the UK's independent fact-checking organization; before that, he was editorial director at BuzzFeed UK. His previous book, *Humans: A Brief History of How We F*cked It All Up* (a history of failure through the ages), was published in 2018 and has sold in 30 territories worldwide. His career has also involved spells of being a comedian, working in television, working in Parliament, and being a secret agent. One of those things is a lie.

ENDNOTES

Introduction: Moment of Truth

1. Kessler, Glenn, Salvador Rizzo, and Meg Kelly, "President Trump Has Made 10,796 False or Misleading Claims over 869 Days," *Washington Post*, June 10, 2019, https://www.washingtonpost.com/politics/2019/06/10/president-trump-has-made-false-or-misleading-claims-over-days/.

2. Kessler, Glenn, "A Year of Unprecedented Deception: Trump Averaged 15 False Claims a Day in 2018," *Washington Post*, December 30, 2018, https://www.washingtonpost.com/politics/2018/12/30/year-unprecedented-deception-trump-averaged-false-claims-day/.

3. Kessler, Glenn, Salvador Rizzo, and Meg Kelly, "President Trump Has Made More Than 5,000 False or Misleading Claims," *Washington Post*, September 13, 2018, https://www.washingtonpost.com/politics/2018/09/13/president-trump-has-made-more-than-false-or-misleading-claims/.

The Origin of the Specious

4. Dekker, Thomas, *The Seven Deadly Sins of London* (Edward Arber, 1879), 21.

5. de Montaigne, Michel, *The Essays of Montaigne, Complete*, Translated by Charles Cotton, Project Gutenberg, 2006, available at https://www.gutenberg.org/files/3600/3600-h/3600-h.htm#link2HCH0009.

6. Machiavelli, Niccolò, "Letter #179, to Franceso Guicciardini, 17 May 1521," quoted in Dallas G. Denery II, *The Devil Wins: A History of Lying from the Garden of Eden to the Enlightenment* (Princeton University Press, 2015), 258.

7. Quoted in Paul V. Trovillo, "History of Lie Detection," *Journal of Criminal Law and Criminology* 29, no. 6 (1938–1939): 849.

8. See Trovillo; also Henry Charles Lea, *Superstition and Force*, 3rd ed., rev. (Henry C. Lea, 1878), 295; and Ali Ibrahim Khan, "On the Trial by Ordeal, among the Hindus," in Sir William Jones, *Supplemental Volumes Containing the Whole of the Asiatick Researches* (G.G. and J. Robinson, 1801), 172.

9. Stefansson, Vilhjalmur, *Adventures in Error* (R. M. McBride & Company, 1936), 7, available at https://hdl.handle.net/2027/wu.89094310885.

10. Sebeok, Thomas A., "Can Animals Lie?" in *I Think I Am a Verb* (Springer, 1986), 128.

11. Angier, Natalie, "A Highly Evolved Propensity for Deceit," *New York Times*, December 22, 2008, https://www.nytimes.com/2008/12/23/science/23angi.html.

12. De Waal, F. B., "Intentional Deception in Primates," *Evolutionary Anthropology Issues News and Reviews* 1, no. 3 (1992): 90.

13. Byrne, Richard W., and Nadia Corp, "Neocortex Size Predicts Deception Rate in Primates," *Proceedings: Biological Sciences* 271, no. 1549 (2004).

14. Talwar, Victoria, "Development of Lying and Cognitive Abilities," in *The Oxford Handbook of Lying*, ed. Jörg Meibauer (Oxford University Press, 2018), 401.

15. Feldman, Robert, *Liar: The Truth About Lying* (Ebury Publishing, 2009), chapter 1, Kindle.

Old Fake News

16. Franklin, Benjamin, *Poor Richard's Almanack and Other Writings* (Dover Publications, 2012), 55.

17. Unlike quite a lot of quotes attributed to Twain, he did at least say something very similar: "I can understand perfectly how the report of my illness got about, I have even heard on good authority that I was

dead. James Ross Clemens, a cousin of mine, was seriously ill two or three weeks ago in London, but is well now. The report of my illness grew out of his illness. The report of my death was an exaggeration. The report of my poverty is harder to deal with." Frank Marshall White, "Mark Twain Amused," *New York Journal*, June 2, 1897, reproduced in *Mark Twain: The Complete Interviews*, ed. Gary Scharnhorst (University of Alabama Press, 2006).

18. "Alan Abel, Satirist Created Campaign to Clothe Animals," *New York Times*, January 2, 1980, 39.

19. "Obituary Disclosed as Hoax," *New York Times*, January 4, 1980, 15.

20. Fox, Margalit, "Alan Abel, Hoaxer Extraordinaire, Is (on Good Authority) Dead at 94," *New York Times*, September 17, 2018, https://www.nytimes.com/2018/09/17/obituaries/alan-abel-dies.html.

21. See, for example, Suzette Smith, "The Day We Thought Jeff Goldblum Died," *Portland Mercury*, June 22, 2016, https://www.portlandmercury.com/The-Jeff-Goldblum-Issue/2016/06/22/18265356/the-day-we-thought-jeff-goldblum-died.

22. Regal, Brian, and Frank J. Esposito, *The Secret History of the Jersey Devil* (Johns Hopkins University Press, 2018), chapter 2, Kindle.

23. "Benjamin Franklin," Wikipedia, s.v. "Benjamin Franklin," https://en.wikipedia.org/wiki/Benjamin_Franklin, accessed February 24, 2019.

24. Stowell, Marion Barber, "American Almanacs and Feuds," *Early American Literature* 9, no. 3 (1975): 276–85, http://www.jstor.org/stable/25070683.

25. Quoted in Stowell, "American Almanacs and Feuds."

26. Franklin, Benjamin, *Poor Richard's Almanack and Other Writings* (Dover Publications, 2013), 28–9.

27. Swift, Jonathan, *Bickerstaff-Partridge Papers* (Amazon Digital Services, 2012), 6, Kindle.

28. Stowell, "American Almanacs and Feuds."

29. Leeds, Titan, quoted in Franklin, *Poor Richard's Almanack*, 30–31.

30. Pettegree, Andrew, *The Invention of News: How the World Came to Know about Itself* (Yale University Press, 2014), 2.

31. Pettegree, *The Invention of News*, 107.

32. Dittmar, Jeremiah, and Skipper Seabold, "Gutenberg's Moving Type Propelled Europe Towards the Scientific Revolution," *LSE Business Review*, March 19, 2019, https://blogs.lse.ac.uk/businessreview/2019/03/19/gutenbergs-moving-type-propelled-europe-towards-the-scientific-revolution/.

33. Schröder, Thomas, "The Origins of the German Press," in *The Politics of Information in Early Modern Europe*, ed. Brendan Dooley and Sabrina A. Baron (Routledge, 2001), 123.

34. Van Groesen, Michiel, "Reading Newspapers in the Dutch Golden Age," *Media History* 22, nos. 3–4 (2016): 336.

35. Schröder, "The Origins of the German Press," 123.

36. Schröder, "The Origins of the German Press," 137.

37. Baillet, Adrien, *Jugemens des sçavans sur les principaux ouvrages des auteurs* (1685), quoted in Ann Blair, "Reading Strategies for Coping with Information Overload ca. 1550–1700," *Journal of the History of Ideas* 64, no. 1 (2003): 11.

38. Burton, *The Anatomy of Melancholy* (EGO Books, 2008), loc. 1337–1347, Kindle.

39. Burton, Robert, *The Anatomy of Melancholy*, loc. 1491–1492, Kindle. The first phrase is in Latin ("Quis tam avidus librorum helluo"); translation per George Hugo Tucker, "Justus Lipsius and the Cento Form," in Erik De Bom et al., *(Un)masking the Realities of Power: Justus Lipsius and the Dynamics of Political Writing in Early Modern Europe* (Brill, 2010), 166.

40. Burton, *The Anatomy of Melancholy*, loc. 1376–1379, Kindle.

41. Dooley, Brendan, "News and Doubt in Early Modern Culture," in Dooley and Baron, *The Politics of Information*, 275.

42. O'Neill, Lindsay, "Dealing with Newsmongers: News, Trust, and Letters in the British World, ca. 1670–1730," *Huntington Library Quarterly* 76, no. 2 (2013): 215–233.

43. Hadfield, Andrew, "News of the Sussex Dragon," in Simon F. Davies and Puck Fletcher, *News in Early Modern Europe—Currents and Connections* (Brill, 2014), 85–86.

44. Quoted in Hadfield, "News of the Sussex Dragon," 88.

45.   Quoted in Markman Ellis, *Eighteenth-Century Coffee-House Culture*, vol. 4 (Routledge, 2006), chapter 6.

46.   James II, "By the King, a Proclamation. To Restrain the Spreading of False News," October 26, 1688, University of Oxford Text Archive, http://tei.it.ox.ac.uk/tcp/Texts-HTML/free/A87/A87488.

47.   From the *Craftsman*, July 17, 1734, quoted in Daniel Woolf, "News, History and the Construction of the Present in Early Modern England," in Dooley and Baron, *The Politics of Information*, 100.

48.   Steiner, Prudence L., "Benjamin Franklin Biblical Hoaxes," in *Proceedings of the American Philosophical Society* 131, no. 2 (1987): 183–196.

The Misinformation Age

49.   "Great Astronomical Discoveries Lately Made by Sir John Herschel, L.L.D. F.R.S. &c. At the Cape of Good Hope [From Supplement to the Edinburgh Journal of Science]," *New York Sun*, August 25, 1835; text from the Museum of Hoaxes, http://hoaxes.org/text/display/the_great_moon_hoax_of_1835_text.

50.   Goodman, Matthew, *The Sun and the Moon: The Remarkable True Account of Hoaxers, Showmen, Dueling Journalists, and Lunar Man-Bats in Nineteenth-Century New York* (Basic Books, 2008), Kindle.

51.   "The Great Moon Hoax," Museum of Hoaxes, http://hoaxes.org/archive/permalink/the_great_moon_hoax.

52.   Griggs, William N., *The Celebrated "Moon Story," Its Origin and Incidents; with a Memoir of the Author, and an Appendix* (Bunnell and Price, 1852), 23–25.

53.   Poe, Edgar Allan, "Richard Adams Locke," in *Complete Works of Edgar Allan Poe* (Delphi Classics, 2015), 1950.

54.   The *Herald*'s name changes go as follows: it was the *Morning Herald* from May to August 1835, became just the *Herald* in late August, and went back to the *Morning Herald* in May 1837, before finally settling on the *New York Herald* in September 1840. See Louis H. Fox, "New York City Newspapers, 1820–1850: A Bibliography," *The Papers of the Bibliographical Society of America* 21, no. 1/2 (1927): 52, http://www.jstor.org/stable/24292637.

55.   Goodman, Matthew, *The Sun and the Moon: The Remarkable True Account of Hoaxers, Showmen, Dueling Journalists, and Lunar Man-Bats in Nineteenth-Century New York* (Basic Books, 2008), Kindle.

56. "The Great Moon Hoax," Museum of Hoaxes, http://hoaxes.org/archive/permalink/the_great_moon_hoax.

57. Phillips, Tom, "25 Things That Will Definitely Happen in the General Election Campaign," BuzzFeed, January 27, 2015, https://www.buzzfeed.com/tomphillips/topless-barry-for-prime-minister.

58. Tucher, Andie, "Those Slippery Snake Stories," *Humanities* 36, no. 3 (May/June 2015), https://www.neh.gov/humanities/2015/mayjune/feature/those-slippery-snake-stories.

59. Tucher, Andie, "The True, the False, and the 'Not Exactly Lying,'" in *Literature and Journalism: Inspirations, Intersections and Inventions from Ben Franklin to Stephen Colbert*, ed. Mark Canada (Palgrave Macmillan, 2013), 91–118.

60. Hills, William H., "Advice to Newspaper Correspondents III: Some Hints on Style," *The Writer*, June 1887, quoted in Tucher, "The True, the False," 93.

61. Hills, William H., "Advice to Newspaper Correspondents IV: Faking," *The Writer*, November 1887, quoted in Tucher, "The True, the False," 93.

62. Shuman, Edwin L., *Steps into Journalism: Helps and Hints for Young Writers* (1894), quoted in Tucher, "The True, the False," 95.

63. MacDougall, Curtis D., *Hoaxes* (Dover Publications, 1958), 4.

64. MacDougall, *Hoaxes*, 4.

65. Khomami, Nadia, "Disco's Saturday Night Fiction," *Observer*, June 26, 2016, https://www.theguardian.com/music/2016/jun/26/lie-heart-disco-nik-cohn-tribal-rites-saturday-night-fever.

66. "Railways and Revolvers in Georgia," *The Times*, October 15, 1856, 9.

67. Untitled article, column 4, "It is assumed by the myriads who sit in judgement…" *The Times*, October 16, 1856, 6.

68. Quoted in E. Merton Coulter, "The Great Georgia Railway Disaster Hoax on the *London Times*," *Georgia Historical Quarterly* 56, no. 1, 1972.

69. Crawford, Martin, "The Great Georgia Railway Disaster Hoax Revisited," *Georgia Historical Quarterly* 58, no. 3, 1974.

70. "The Southern States of America," *The Times*, August 27, 1857, 8.

71. "Comet's Poisonous Tail," *New York Times*, February 8, 1910, 1.

72. "Some Driven to Suicide," *New York Times*, May 19, 1910, 2.

73. Alexander, Stian, "Croydon Cat Killer Has Widened Brutal Spree around the M25, Say Police," *Daily Mirror*, July 13, 2016, https://www.mirror.co.uk/news/uk-news/croydon-cat-killer-widened-brutal-8414154.

74. Smith, Patrick, "The Met Police Spent More Than 2,000 Hours Investigating The 'Croydon Cat Killer,'" BuzzFeed, December 18, 2018, https://www.buzzfeed.com/patricksmith/operation-takahe-costs-police-hours.

75. "Mattoon Gets Jitters from Gas Attacks," *Chicago Herald-American*, September 10, 1944, quoted in Robert Bartholomew and Hilary Evans, *Panic Attacks: Media Manipulation and Mass Delusion* (The History Press, 2004).

76. "On the Contrary," *New Yorker*, December 9, 2002, https://www.newyorker.com/magazine/2002/12/09/on-the-contrary.

77. https://twitter.com/baltimoresun/status/1028118771192528897—the wall gave a date of 1953; it was actually from 1946.

78. Mencken, H. L., "Melancholy Reflections," *Chicago Tribune*, May 23, 1926, 74.

79. Mencken, "Melancholy Reflections," 74.

80. Mencken, H. L., "A Neglected Anniversary," *New York Evening Mail*, December 28, 1917.

81. Stefansson, *Adventures in Error*, 288–90.

82. Hersey, John, "Mr. President IV: Ghosts in the White House," *New Yorker*, April 28, 1951, 44–45, https://www.newyorker.com/magazine/1951/04/28/mr-president-ghosts-in-the-white-house.

83. Truman, Harry S., "Address in Philadelphia at the American Hospital Association Convention," September 16, 1952, Harry S. Truman Presidential Library & Museum, https://www.trumanlibrary.gov/library/public-papers.

84. Fleischman, Sandra, "Builders' Winning Play: A Royal Flush," *Washington Post*, November 24, 2001; and Andrea Sachs, "Presidents' Day 101," *Washington Post*, February 15, 2004.

85. Mencken, H. L., "Hymn to the Truth," *Chicago Tribune*, July 25, 1926, 61.

The Lie of the Land

86. Burton, R. F., "The Kong Mountains," *Proceedings of the Royal Geographical Society and Monthly Record of Geography* 4, no. 8 (1882): 484–86, https://www.jstor.org/stable/1800716.

87. There are only two maps known to have the Latin phrase *Hic sunt dracones* on them, both from the early 1500s. The phrase is never known to have appeared in English. See Meeri Kim, "Oldest Globe to Depict the New World May Have Been Discovered," *Washington Post*, August 19, 2013, https://www.washingtonpost.com/national/health-science/oldest-globe-to-depict-the-new-world-may-have-been-discovered/2013/08/19/503b2b4a-06b4-11e3-a07f-49ddc7417125_story.html.

88. Rennell, James, "A Map, Shewing the Progress of Discovery & Improvement, in the Geography of North Africa," 1798, Library of Congress Geography and Map Division, Washington, DC, https://www.loc.gov/item/2009583841/.

89. Bassett, Thomas J., and Philip W. Porter, "'From the Best Authorities': The Mountains of Kong in the Cartography of West Africa," *Journal of African History* 32, no. 3 (1991): 367–413, www.jstor.org/stable/182661.

90. Park, Mungo, *Life and Travels of Mungo Park in Central Africa* (Amazon Digital Services, 2012), 181, Kindle.

91. Rennell, James, *Proceedings of the Association for Promoting the Discovery of the Interior Parts of Africa* (W. Bulmer & Co, 1798), 63.

92. Brooke-Hitching, Edward, *The Phantom Atlas: The Greatest Myths, Lies and Blunders on Maps* (Simon & Schuster UK, 2016).

93. Burton, "The Kong Mountains," 484–86.

94. Clapperton, Hugh, Richard Lander, and Abraham V. Salamé, *Journal of a Second Expedition into the Interior of Africa, from the Bight of Benin to Soccatoo* (John Murray, 1829), 21.

95. Bassett, Thomas J., and Philip W. Porter, "'From the Best Authorities': The Mountains of Kong in the Cartography of West Africa," in *The Journal of African History* 32, no. 3 (1991): 373.

96. Binger, Louis-Gustave, "Du Niger au Golfe de Guinee par Kong," in Bulletin de la Société de Géographie (Paris), 1889, quoted in Bassett and Porter, "'From the Best Authorities': The Mountains of Kong in

the Cartography of West Africa," in *The Journal of African History* 32, no. 3 (1991): 395.

97. Adams, Percy G., *Travelers and Travel Liars 1660–1800* (Dover Publications, 1980), 158–161.

98. Brooke-Hitching, Edward, *The Phantom Atlas: The Greatest Myths, Lies and Blunders on Maps* (Simon & Schuster UK, 2016), 166.

99. Campbell, Matthew, "Oil Boom Fuels Mystery of the Missing Island in the Mexican Gulf," *The Times*, September 6, 2009, https://www.thetimes.co.uk/article/oil-boom-fuels-mystery-of-the-missing-island-in-the-mexican-gulf-xg7tcsdbcwz.

100. "How, Modestly, Cook Hoaxed the World," *New York Times*, December 22, 1909, 4, https://www.nytimes.com/1909/12/22/archives/how-modestly-cook-hoaxed-the-world-turned-a-smiling-face-to-critics.html.

The Scam Manifesto

101. "The King of Con-Men," *Economist*, December 22, 2012, https://www.economist.com/christmas-specials/2012/12/22/the-king-of-con-men.

102. "On the first day March, 1822, the price will be advanced One Shilling and Sixpence per Acre, and in the same proportion every three months hereafter." From "North America," *Perthshire Courier*, December 20, 1821, 1.

103. *The Times*, July 12, 1822, 1.

104. Strangeways, Thomas, *Sketch of the Mosquito Shore, Including the Territory of Poyais* (1822).

105. Conzemius, Eduard, "Ethnographical Survey of the Miskito and Sumu Indians of Honduras and Nicaragua," *Bureau of American Ethnology Bulletin* 106 (1932): 1; quoted in V. Wolfgang von Hagen, "The Mosquito Coast of Honduras and Its Inhabitants," *Geographical Review* 30, no. 2 (1940): 252.

106. Compare the map in *Sketch of the Mosquito Coast* with both modern maps and the map in Hagen, "The Mosquito Coast," 240.

107. Raista Eco Lodge is apparently "community based tourism at its best": "Laguna de Ibans," Lonely Planet, https://www.lonelyplanet.

com/honduras/hotels/raista-eco-lodge/a/lod/9c119bd1-1ee4-4e60-8a53-e9ba4279a71e/1328476.

108. *Manchester Guardian*, October 25, 1823. Republished as "Settlers Duped into Believing in 'Land Flowing with Milk and Honey,'" *Guardian*, October 25, 2013, https://www.theguardian.com/the-guardian/2013/oct/25/gregor-macgregor-poyais-settlers-scam.

109. Rafter, Michael, *Memoirs of Gregor M'Gregor: Comprising a Sketch of the Revolution in New Grenada and Venezuela* [...] (J. J. Stockdale, 1820), 19.

110. Brown, Matthew, "Inca, Sailor, Soldier, King: Gregor MacGregor and the Early Nineteenth-Century Caribbean," *Bulletin of Latin American Research* 24, no. 1 (2005): 55.

111. Rafter, *Memoirs of Gregor M'Gregor*, 20.

112. Rafter, *Memoirs of Gregor M'Gregor*, 19.

113. Weatherhead, W. D., *An Account of the Late Expedition against the Isthmus of Darien under the Command of Sir Gregor M'Gregor* (Longman, Hurst, Rees, Orme, and Brown, 1821), 26.

114. *Jamaica Gazette*, July 17, 1819, quoted in Brown, "Inca, Sailor, Soldier, King," 59.

115. Rafter, *Memoirs of Gregor M'Gregor*, 338.

116. For more of this, see Brown, "Inca, Sailor, Soldier, King."

117. *London Literary Gazette and Journal of Belles Lettres, Arts, Sciences, Etc.* 315, February 1 (1823): 70.

118. Possibly worth quoting this at length: "The savage criticism on his *Endymion*, which appeared in the *Quarterly Review*, produced the most violent effect on his susceptible mind; the agitation thus originated ended in the rupture of a blood-vessel in the lungs; a rapid consumption ensued, and the succeeding acknowledgments from more candid critics, of the true greatness of his powers, were ineffectual to heal the wound thus wantonly inflicted." Percy B. Shelley, preface to *Adonais: An Elegy on the Death of John Keats, Author of Endymion, Hyperion, Etc.* (1821).

119. "Art. VIII," in *Quarterly Review* 28 (October 1822–January 1823): 157–61.

120. Frankel, Tamar, *The Ponzi Scheme Puzzle* (Oxford University Press, 2012), 111.

121. Frankel, *The Ponzi Scheme Puzzle*, 89.

122. Frankel, *The Ponzi Scheme Puzzle*, 85.

123. Konnikova, Maria, *The Confidence Game: The Psychology of the Con and Why We Fall for It Every Time* (Canongate Books, 2016), 8.

124. Kerenyi, Norbert, *Stories of a Survivor* (Xlibris, 2011), 280.

125. McCarthy, Joe, "The Master Impostor: An Incredible Tale," *Life*, January 28, 1952, 81.

126. "'Master Impostor' Now May Try to Be Just Himself," *Minneapolis Sunday Tribune*, January 8, 1956, 10A.

127. Associated Press, "Ferdinand Waldo Demara, 60, an Impostor in Varied Fields," *New York Times*, June 9, 1982, B16.

128. Crichton, Robert, *The Great Impostor* (Random House, 1959), 103.

129. Alexopoulos, Golfo, "Portrait of a Con Artist as a Soviet Man," *Slavic Review* 57, no. 4 (Winter, 1998): 775.

130. Zaleski, Eugène, *Stalinist Planning for Economic Growth, 1933–1952* (University of North Carolina Press, 1980), quoted in Alexopoulos, "Portrait of a Con Artist," 777.

131. Alexopoulos, "Portrait of a Con Artist," 781.

132. Quoted in Alexopoulos, "Portrait of a Con Artist," 788.

133. Spurling, Hilary, *La Grande Thérèse: The Greatest Swindle of the Century* (Profile Books, 2000), 24.

134. Quoted in Spurling, *La Grande Thérèse*, 44.

135. Martin, Benjamin F., *The Hypocrisy of Justice in the Belle Epoque* (Louisiana State University Press, 1984), 80.

136. Quoted in Spurling, *La Grande Thérèse*, 48.

Lying in State

137. Almond, Cuthbert, "Oates's Plot," *Catholic Encyclopedia*, https:// www. catholic.com/encyclopedia/oatess-plot.

138. Marshall, Alan, "Titus Oates," Oxford Dictionary of National Biog-

raphy, January 3, 2008, https://www.oxforddnb.com/view/10.1093/ref:odnb/9780198614128.001.0001/odnb-9780198614128-e-20437.

139. Pollock, Sir John, *The Popish Plot: A Study in the History of the Reign of Charles II* (Duckworth & Co., 1903), 3.

140. Kopel, David, "The Missing 18 1/2 Minutes: Presidential Destruction of Incriminating Evidence," *Washington Post*, June 16, 2014, https://www.washingtonpost.com/news/volokh-conspiracy/wp/2014/06/16/the-missing-18-12-minutes-presidential-destruction-of-incriminating-evidence/.

141. McDonald, Iverach, *The History of the Times: Volume V, Struggles in Life and Peace, 1939–1966* (Times Books, 1984), 268–269.

142. "Crucifixion of Canadians (Alleged)," *Hansard*, May 19, 1915, https://api.parliament.uk/historic-hansard/commons/1915/may/19/crucifixion-of-canadians-alleged#S5CV0071P0-08398.

143. "Through German Eyes," *The Times*, April 16, 1917, 7.

144. "Supplement to the Boston *Independent Chronicle*" [before April 22, 1782], Founders Online, National Archives, last modified June 13, 2018, http://founders.archives.gov/documents/Franklin/01-37-02-0132. Original source: *The Papers of Benjamin Franklin Volume 37: March 16 through August 15, 1782*, ed. Ellen R. Cohn (Yale University Press, 2003), 184–196.

145. Mulford, Carla, "Benjamin Franklin's Savage Eloquence: Hoaxes from the Press at Passy, 1782," *Proceedings of the Amercian Philosophical Society* 152, no. 4 (2008): 497.

146. "I send enclosed a Paper, of the Veracity of which I have some doubt, as to the Form, but none as to the Substance, for I believe the Number of People actually scalp'd in this murdering War by the Indians to exceed what is mention'd in the Invoice." See "From Benjamin Franklin to John Adams, 22 April 1782," Founders Online, National Archives, last modified June 13, 2018, http://founders.archives.gov/documents/Franklin/01-37-02-0133. Original source: *The Papers of Benjamin Franklin Volume 37*, 196–97.

147. Dowd, Gregory Evans, *Groundless: Rumors, Legends and Hoaxes on the Early American Frontier* (Johns Hopkins University Press, 2016), 170–172.

Funny Business

148. Manes, Stephen, *Gates: How Microsoft's Mogul Reinvented an Industry—and Made Himself the Richest Man in America* (Cadwallader & Stern, 1993), chapter 5, Kindle.

149. Merchant, Brian, *The One Device: The Secret History of the iPhone* (Bantam Press, 2017), 367. It's rather noticeable, when you watch the video of the launch, that, when Jobs makes staged phone calls to Jony Ive and Phil Schiller, neither of them is actually using an iPhone—they both have old-school flip phones. See YouTube video, 25:34, https://www.youtube.com/watch?v=9hUIxyE2Ns8.

150. MacRory, Henry, *Ultimate Folly: The Rises and Falls of Whitaker Wright* (Biteback Publishing, 2018), chapter 7, Kindle.

151. Quoted in MacRory, *Ultimate Folly*, chapter 3, Kindle.

152. Quoted in MacRory, *Ultimate Folly*, chapter 2, Kindle.

153. Mackay, Charles, *Memoirs of Extraordinary Popular Delusions and the Madness of Crowds*, 1852, Kindle.

154. Oppenheim, A. Leo, *Letters from Mesopotamia* (University of Chicago Press, 1967), 82–83.

155. All quotes in this passage are from Michael Rice, *The Archaeology of the Arabian Gulf* (Routledge, 2002), 276–78.

156. Levi, Steven C., "P. T. Barnum and the Feejee Mermaid," in *Western Folklore* 36, no. 2 (1977): 149–54.

157. Reiss, Benjamin, "P. T. Barnum, Joice Heth and Antebellum Spectacles of Race," in *American Quarterly* 51, no. 1 (1999): 78–107.

158. Lee, R. Alton, *The Bizarre Careers of John R. Brinkley* (University Press of Kentucky, 2002), Kindle.

Ordinary Popular Delusions

159. Rowlatt, Justin, "Gatwick Drone Attack Possible Inside Job, Say Police," BBC News, April 14, 2019, https://www.bbc.co.uk/news/uk-47919680.

160. "Gatwick Drones Pair 'No Longer Suspects,'" BBC News, December 23, 2018, http://web.archive.org/web/20181223172230/https://

www.bbc.co.uk/news/uk-england-46665615. The BBC subsequently
edited the article to remove the police quote.

161. Rowlatt, "Gatwick Drone Attack."

162. See the interactive map at Brett Holman, "Mapping the 1913 Phantom
Airship Scare," *Airminded* (blog), https://airminded.org/2013/05/03/
mapping-the-1913-phantom-airship-scare/.

163. Hirst, Francis Wrigley, *The Six Panics and Other Essays* (Methuen,
1913), 104.

164. Bartholomew, Robert E., *Hoaxes, Myths, and Manias: Why We Need
Critical Thinking* (Prometheus Books, 2003), chapter 9, Kindle.

165. Mattalaer, Johan J. and Wolfgang Jilek "Koro—The Psychological
Disappearance of the Penis," *Journal of Sexual Medicine* 4, no. 5 (2007).

166. Bartholomew, Robert E., *A Colorful History of Popular Delusions*
(Prometheus Books, 2015), 37.

167. Barzilay, Tzafrir, "Well-Poisoning Accusations in Medieval Europe:
1250–1500" (doctoral thesis, Columbia University, 2016), 95, https://
academiccommons.columbia.edu/doi/10.7916/D8VH5P6T.

168. Gui, Bernard, *Vita Joannis* XXII, 163, quoted in Barzilay, "Well-
Poisoning," 110.

169. Leeson, P. T., and J. W. Russ, "Witch Trials," *Economic Journal* 128,
no. 613 (2018).

170. "Posse Sets Out as 'Jersey Devil' Reappears," *New York Times*, December 19, 1929, 14.

171. "Posse Sets Out as 'Jersey Devil' Reappears," *New York Times*, December 19, 1929, 14.

172. Regal, Brian, and Frank J. Esposito, *The Secret History of the Jersey
Devil* (Johns Hopkins University Press, 2018), Kindle.

Conclusion: Toward a Truthier Future

173. Breves, Dylan, "Coati," Wikipedia, revision as of 02.36 UTC, 12
July 2008, https://en.wikipedia.org/w/index.php?title=Coati&diff=
next&oldid=224679361, accessed June 29, 2019. (Previous absence of
the term as per results of date-limited searches on Google, Google
Scholar and Google Books.)

174. Randall, Eric, "How a Racoon Became an Aardvark," *New Yorker,* May 19, 2014, https://www.newyorker.com/tech/annals-of-technology/how-a-raccoon-became-an-aardvark; Sightings in the press include: Williams, Amanda, "Hunt for the runaway aardvark: Lady McAlpine calls on public to help find her lost ring-tailed coati," *Daily Mail,* April 8, 2013, https://www.dailymail.co.uk/news/article-2305602/Hunt-runaway-aardvark-LadyMcAlpine-calls-public-help-lost-ring-tailed-coati.html; Leach, Ben, "Scorpions, Brazilian aardvarks and wallabies all found living wild in UK, study finds," *Daily Telegraph,* June 21, 2010, https://www.telegraph.co.uk/news/earth/wildlife/7841796/Scorpions-Brazilianaardvarks-and-wallabies-all-found-living-wild-in-UK-study-finds.html; Brown, Jonathan, "From wallabies to chipmunks, the exotic creatures thriving in the UK," *Independent,* June 21, 2010, https://www.independent.co.uk/environment/nature/from-wallabies-tochipmunks-the-exotic-creatures-thriving-in-the-uk-2006096.html (although this reference is only to "aardvarks").

175. "Scorpions and Parakeets 'Found Living Wild in UK,'" BBC News, June 21, 2010, https://www.bbc.co.uk/news/10365422. You'll notice that several of these are versions of the same story, about non-native species living wild in the UK. These were all based on a "report" by a University of Hull academic, commissioned for PR purposes by Eden, a television channel; it seems likely that the mistake was included in the original press release and copied from there, although I haven't been able to track down the original press release.

176. Nadal, James, "Brazilian Aardvark on the Loose in Marlow," *Bucks Free Press,* February 20, 2013, https://www.bucksfreepress.co.uk/news/10240842.brazilian-aardvark-on-the-loose-in-marlow/; Drury, Flora, "So That's What an Aardvark Looks Like," *Worcester News,* June 9, 2011, https://www.worcesternews.co.uk/news/9072841.so-thats-what-an-aardvark-looks-like/.

177. "Photo of the Day: Wild Fire," *Time,* September 20, 2013, https://time.com/3802583/wild-fire/; "An Unexpected Visitor in the Volcano," *National Geographic,* March 7, 2013, https://blog.nationalgeographic.org/2013/03/07/an-unexpected-visitor-in-the-volcano/; Platt, John R., "Brazil Plans to Clone Its Endangered Species," Extinction Countdown, *Scientific American,* November 14, 2012, https://blogs.scientificamerican.com/extinction-countdown/brazil-plans-to-clone-its-endangered-species/.

178. Cançado, Paulo Henrique Duarte, et al., "Current Status of Ticks and Tick-Host Relationship in Domestic and Wild Animals from Pan-

tanal Wetlands in the State of Mato Grosso do Sul, Brazil," *Iheringia. Série Zoologia* 107, supplement (May 2, 2017) https://dx.doi.org/10.1590/1678-4766e2017110.

179. Henderson, Caspar, *The Book of Barely Imagined Beings: A 21st Century Bestiary* (University of Chicago Press, 2013), 10.

180. Safier, Neil, "Beyond Brazilian Nature: The Editorial Itineraries of Marcgraf and Piso's Historia Naturalis Brasiliae," in *The Legacy of Dutch Brazil*, ed. Michiel van Groesen (Cambridge University Press, 2014), 179, https://doi.org/10.1017/CBO9781107447776.011.

181. "David Attenborough and BBC Take Us to Hotel Armadillo—in Pictures," *Guardian*, April 5, 2017, https://www.theguardian.com/environment/gallery/2017/apr/05/david-attenborough-and-bbc-take-us-to-hotel-armadillo-in-pictures.

182. Kolbe, Andreas, "Happy birthday: Jimbo Wales' sweet 16 Wikipedia fails", *The Register*, January 16, 2017, https://www.theregister.co.uk/2017/01/16/wikipedia_16_birthday_fails/.

183. Allen, Nick, "Wikipedia, the 25-Year-Old Student and the Prank that Fooled Leveson," *Daily Telegraph*, December 5, 2012, https://www.telegraph.co.uk/news/uknews/leveson-inquiry/9723296/Wikipedia-the-25-year-old-student-and-the-prank-that-fooled-Leveson.html.

184. Wikipedia, s.v. "Wikipedia: List of citogenesis incidents," accessed June 30, 2019, https://en.wikipedia.org/wiki/Wikipedia:List_of_citogenesis_incidents.

185. Phillips, Tom, *Truth: A Brief History of Total Bullshit* (Wildfire, 2019), 19.

186. Dash, Mike, "The Origin of the Tale that Gavrilo Princip Was Eating a Sandwich When He Assassinated Franz Ferdinand," *Smithsonian Magazine*, September 15, 2011, https://www.smithsonianmag.com/history/gavrilo-princips-sandwich-79480741/.

187. Stefansson, Vilhjalmur, *Adventures in Error* (R. M. McBride & Company, 1936), 16, available at https://hdl.handle.net/2027/wu.89094310885.

188. Franklin, Benjamin, "Report of Dr. Benjamin Franklin, and other commissioners, charged by the King of France, with the examination of the animal magnetism, as now practised at Paris," p. xvii, Wellcome Library, available at https://wellcomelibrary.org/item/b20595244.

INDEX

Page numbers in *italics* indicate illustrations.

manias, 26, 243, 248, 262–263, 277–278. *See also* witch hunts

maps, misinformation on, 25. *See also* Mountains of Kong

marketers, 22

Mark Inside, The (Reading), 277

mass hallucination, 241

mass hysteria, 26, 39, 112–113, 192–194

mass media
lies and, 22
origins of, 66–67, 81
PT Barnum and, 227
scams and, 168–169
See also lies, cons, delusions, hoaxes: specific; news industry

Mattoon Daily Journal-Gazette, 111–112

medicine, fake, 227–231, 277

Mencken, H. L., 113–119, 201

mendacity, spontaneous, 95, 115

merchant class, 67

Mesmer, Anton, 230–231, 267

Microsoft, 214, 215

mining scam, 218–219

Miranda, Francisco de, 162

misinformation, 22, 25, 26, 43–44, 98, 257–262, 264

Montaigne, Michel de, 34, 267–268

moral panics, 26, 243

Morning Herald (New York Herald), 93, 96, 167, 293n54

Morrell, Benjamin, 138

Morrell's Island, 138

Mosquito Coast, 156

motivated reasoning, 50–51

Mountains of Kong, 125–136, *128, 134*

Mountains of the Moon, 126, 131–133, *132, 134*, 137

Murdoch, Rupert, 90

Museum of Hoaxes, 97

mythical lands, 25, 138–145, 151–152, 154–157, 164–165, 167–168. *See also* Mountains of Kong; Mountains of the Moon; Poyais scheme

Myths & Legends of the First World War (Hayward), 277

Nanni, 222–223

National Geographic Society, 139, 141

Native Americans, 207–208

natural theology, 97

networks, postal, 67, 68

Neuigkeitssucht, 70–71

news
addiction to, 70–71
distrust of, 73
fake (*See* fake news)
republishing of, 81
truth in, 100–101

news industry
almanacs, 61–63
Benjamin Franklin and (*See* Franklin, Benjamin)
development and distribution of newspapers, 24, 69–70, 90–92, 97
Great Moon Hoax of 1835, 24, *89*, 89–98, 113, 276